ARSENIC: NATURAL AND ANTHROPOGENIC

Arsenic in the Environment

Series Editors

Jochen Bundschuh

University of Applied Sciences, Institute of Applied Research, Karlsruhe, Germany
Royal Institute of Technology (KTH), Stockholm, Sweden

Prosun Bhattacharya

KTH-International Groundwater Arsenic Research Group, Department of Land and Water Resources Engineering, Royal Institute of Technology (KTH), Stockholm, Sweden

ISSN: 1876-6218

Volume 4

ISGSD

International Society of
Groundwater for
Sustainable Development

Cover photo

The cover photo of volume 4 shows a scene from the human biomonitoring (HUBI) performed in the ARSENEX project in Brazil.

Visible are tubes for the creatinine sub-samples (top), the questionnaires for the participating children and adults (center), and prepared sample stickers (bottom).

Arsenic: Natural and anthropogenic

Editors

Eleonora Deschamps
Fundação Estadual do Meio Ambiente, Belo Horizonte, MG, Brazil

Jörg Matschullat
Technische Universität Bergakademie Freiberg, Germany

CRC Press
Taylor & Francis Group
Boca Raton London New York

CRC Press is an imprint of the
Taylor & Francis Group, an **informa** business

A BALKEMA BOOK

CRC Press
Taylor & Francis Group
6000 Broken Sound Parkway NW, Suite 300
Boca Raton, FL 33487-2742

First issued in paperback 2018

CRC Press/Balkema is an imprint of the Taylor & Francis Group, an informa business

© 2011 Taylor & Francis Group, LLC

ISBN-13: 978-0-415-54928-8 (hbk)
ISBN-13: 978-1-138-07310-4 (pbk)

Typeset by Vikatan Publishing Solutions (P) Ltd., Chennai, India

Published by: CRC Press/Balkema
　　　　　P.O. Box 447, 2300 AK Leiden, The Netherlands
　　　　　e-mail: Pub.NL@taylorandfrancis.com
　　　　　www.crcpress.com – www.taylorandfrancis.co.uk – www.balkema.nl

Library of Congress Cataloging-in-Publication Data

　Arsenic : natural and anthropogenic / editors, Eleonora Deschamps, Jörg Matschullat.
　　　p. cm. -- (Arsenic in the environment ; 4)
　ISBN 978-0-415-54928-8 (hardback) -- ISBN 978-0-203-09322-1 (ebook) 1. Arsenic--Environmental aspects. I. Matschullat, Jörg. II. Deschamps, Eleonora. III. Title. IV. Series.

　TD427.A77A7655 2011
　628.5'2--dc22

　　　　　　　　　　　　　　　　　　　2011001240

Visit the Taylor & Francis Web site at
http://www.taylorandfrancis.com

and the CRC Press Web site at
http://www.crcpress.com

About the book series

Although arsenic has been known as a 'silent toxin' since ancient times, and the contamination of drinking water resources by geogenic arsenic was described in different locations around the world long ago – e.g., in Argentina in 1917 – it was only two decades ago that it received overwhelming worldwide public attention. As a consequence of the biggest arsenic calamity in the world, which was detected more than twenty years back in Bangladesh, West Bengal, India and other parts of Southeast Asia, there has been an exponential rise in scientific interest that has triggered high quality research. Since then, arsenic contamination (predominantly of geogenic origin) of drinking water resources, soils, plants and air, the propagation of arsenic in the food chain, the chronic effects of arsenic ingestion by humans, and their toxicological and related public health consequences, have been described in many parts of the world, and every year, even more new countries or regions are discovered to have arsenic problems.

Arsenic is found as a drinking water contaminant in many regions all around the world, in both developing as well as industrialized countries. However, addressing the problem requires different approaches which take into account the different economic and social conditions in both country groups. It has been estimated that 200 million people worldwide are at risk from drinking water containing high concentrations of arsenic, a number which is expected to further increase due to the recent lowering of the limits of arsenic concentration in drinking water to 10 μg L^{-1}, which has already been adopted by many countries, and some authorities are even considering decreasing this value further.

The book series "Arsenic in the Environment" is an inter- and multidisciplinary source of information, making an effort to link the occurrence of geogenic arsenic in different environments and the potential contamination of ground and surface water, soil and air and their effect on the human society. The series fulfills the growing interest in the worldwide arsenic issue, which is being accompanied by stronger regulations on the permissible Maximum Contaminant Levels (MCL) of arsenic in drinking water and food, which are being adopted not only by the industrialized countries, but increasingly by developing countries.

The book series covers all fields of research concerning arsenic in the environment and aims to present an integrated approach from its occurrence in rocks and mobilization into the ground- and surface water, soil and air, its transport therein, and the pathways of arsenic introduction into the food chain including uptake by humans. Human arsenic exposure, arsenic bioavailability, metabolism and toxicology are treated together with related public health effects and risk assessments in order to better manage the contaminated land and aquatic environments and to reduce human arsenic exposure. Arsenic removal technologies and other methodologies to mitigate the arsenic problem are addressed not only from the technological perspective, but also from an economic and social point of view. Only such inter- and multidisciplinary approaches, will allow case-specific selection of optimal mitigation measures for each specific arsenic problem and provide the local population with safe drinking water, food, and air.

We have the ambition to make this book series an international, multi- and interdisciplinary source of knowledge and a platform for arsenic research oriented to the direct solution of problems with considerable social impact and relevance rather than simply focusing on cutting edge and breakthrough research in physical, chemical, toxicological and medical sciences. The book series will also form a consolidated source of information on the worldwide occurrences of arsenic, which otherwise is dispersed and often hard to access. It will also have role in increasing the awareness and knowledge of the arsenic problem among administrators, policy makers and company executives and in improving international and bilateral cooperation on arsenic contamination and its effects.

Consequently, we see this book series as a comprehensive information base, which includes authored or edited books from world-leading scientists on their specific field of arsenic research, but also contains volumes with selected papers from international or regional congresses or other scientific events. Further, the abstracts presented during the homonymous biannual international congress series, which we organize in different parts of the world is being compiled in a stand-alone book series "Arsenic in the Environment – Proceedings" that would give short and crisp state of the art periodic updates of the contemporary trends in arsenic-related research. Both series are open for any person, scientific association, society or scientific network, for the submission of new book projects. Supported by a strong multi-disciplinary editorial board, book proposals and manuscripts are peer reviewed and evaluated.

Jochen Bundschuh
Prosun Bhattacharya
(*Series Editors*)

Editorial board

Table of contents

Foreword

Environmental problems and related human misconduct, deliberate or accidental, fill entire volumes. The ever-increasing awareness of global change issues subsumes those mostly localized problems. Yet, it generates in many the impression that a doomsday scenario would better describe reality and our future than any modest and (self) critical attempt to assess related problems – and to find practical and sustainable solutions. People involved, scientists as much as administrators, need to have a vision – and the courage to step (at least once in a while) beyond the familiar boundaries of each specialists' realm. Such courage and responsibility from all sides is needed if existing problems are not only to be described and explained, but meaningful solutions are to be developed – and tested. The opportunity to pursue such scientific interests, to open new horizons – and at the same time being able to contribute to social development and to assist needy communities is a rare and most gratifying experience.

It was exactly this kind of prospect that initially motivated a handful of scientists (from the State University of Campinas, UNICAMP, in São Paulo; the University of Heidelberg, later TU Bergakademie Freiberg, TUBAF; FEAM, the State Environmental Agency of Minas Gerais, FUNED, the State Health Agency, COPASA, the State Water Agency and the Federal University of Minas Gerais, UFMG, in Belo Horizonte) to begin an endeavour that evolved into the ARSENEX project; the backbone of this volume. A venture of such complexity and far-reaching consequences can neither be launched nor performed successfully without substantial support from various individuals and institutions. Having had the pleasure to be part of this unique project from the beginning, we appreciate the opportunity to express our satisfaction and gratitude to all those who contributed in various ways to the accomplishment of this project, and to the successful improvement of local living conditions.

Figure 0.1.   Towards new horizons (children in their schoolyard, 2001; photo Jörg Matschullat).

We thankfully acknowledge financial and moral support from the German Ministry of Science and Technology (BMBF), the Brazilian Ministry of the Environment (MMA) and the Brazilian National Environment Fund (FNMA), the Brazilian Research Support Foundation of Minas Gerais (FAPEMIG), the German Academic Exchange Service (DAAD), the German Science Foundation (DFG) and FEAM, all representing state and federal ministries and thus the governments.

Without the dedication of the school teachers and principals, and the citizens of the villages and towns in the Nova Lima and Santa Bárbara districts of Minas Gerais, this project would not have been successful either. As a representative of these wonderful people, we like to name Dilce Amara Margarida Mendes from Brumal – and give a big hug to all our friends in the region. At the same time, we thank the late Prof. Dr. Dr. h.c. mult. German Müller (†), Heidelberg University, for his initiative to welcome a Brazilian doctoral student in 1996 to the then scientific home of Jörg Matschullat. German Müller's wide international scope and the enlightening atmosphere in the former Institute of Environmental Geochemistry substantially contributed to this and many other projects. The most dedicated support by the human toxicology group of the Baden Württemberg State Health Agency (*Landesgesundheitsamt*), namely Prof. Dr. Michael Schwenk and Dr. Thomas Gabrio, and their team in Stuttgart, cannot be applauded enough. Their engagement enabled not only the initial urine sampling (▶ 14) and subsequent analysis of those samples from hundreds of children and adults, but also significantly contributed to successfully establishing a similar laboratory infrastructure and staff training at the Minas Gerais State Health Agency (FUNED) in Belo Horizonte.

We are particularly indepted to the Minas Gerais State Environmental Agency (FEAM) and its then and current staff, for constant and substantial support. Without FEAM, the Brazilian edition of a first book on the ARSENEX project would not have come true nor the ongoing support for this volume. A word of appreciation and respect also goes to the local mining companies, namely AngloGold in Nova Lima and São Bento Mineração, today also Anglo Gold, in Barão de Cocais. They bore with us throughout the years, and their – sometimes not easy – commitment to improve the environmental situation even beyond their proper operations became part of the project success.

Many ambitious and engaged students from Brazil and Germany dedicated their minds and thesis work to the topics of this project, and helped compile "building stones" without which the challenges would not have been met in due course. Finally, we wish to warmly and heartfully thank our families and friends for supporting our endeavour. Their understanding provided the freedom to stay out in the field for weeks and sometimes in another country for extended periods of time.

We thank the series editors, Jochen Bundschuh and Prosun Bhattacharya for inviting us to publish this book in the "Arsenic in the environment" series, and the publishers at CRC Press and Balkema, namely Janjaap Blom and his team, for believing in this project. Last, but not least, a warm "thank you" goes to the de Grosbois[3] for language editing, and to Silvia Leise, Kirk Nordstrom, Cornelius Oertel, Alexander Pleßow, Clemens Reimann, Annika Seidler and Frank Zimmermann for their thorough and most helpful additional review of individual chapters.

Some technical remarks. This book is organized in four parts with 17 chapters. The editors took great care to deliver a "book in one piece", despite the multi-author contributions. Cross-references (▶) lead the reader to related material in other book chapters. True references are always listed at the end of each chapter, including additional reading and web-based addresses when it is deemed helpful. The editors made sure that all references were up-to-date and functional when submitting the manuscripts. The element name "arsenic" usually appears abbreviated (As), unless that would cause misunderstandings.

Section I delivers state-of-the-art information on arsenic in the environment (▶ 1), its toxicology (▶ 2), As remediation methods (▶ 3, 4), and related environmental legislation (▶ 5).

Section II introduces the regional background of the ARSENEX project to those, who may not be familiar with that beautiful part of the world (▶ 6, 7). Project philosophy and related background information are given (▶ 8), and the results of a very important study on environmental and health perceptions of people in the target region are presented (▶ 9).

Section III discusses the environmental compartments that needed to be assessed to develop sustainable solutions for the As problem, namely atmosphere, hydrosphere, soils and sediments, edible and bioaccumulating plants, and the human body (▶ 10–14).

Section IV describes important aspects of project consequences, the environmental and health education (▶ 15), the construction of a decentralized water purification plant with citizen participation (▶ 16) and finally some concluding remarks and suggestions that may help others who look for answers to pressing questions, even under restricted financial conditions (▶ 17). This is followed by a subject index that may help in additional cross-referencing.

Outlook. Looking forward to future fruitful collaboration between people and institutions from Brazil and Germany, and to contribute to scientific advancement and a better quality of life for those people who need it most, we thank you, the reader for your interest in this work. You will discover that this is not a "run-of-the-mill" project account. Instead, you hold a book that takes a holistic view at the complex challenge of As contamination triggered by gold mining activities. Broadly similar constellations exist all over the world, related to industry, mining, and natural anomalies, and beyond arsenic for many other potential environmental contaminants. It is for such challenges that we wrote this book to assist others in making not only realiable assessments, but to successfully encounter solutions and minimize or even eliminate environmental risks.

Eleonora Deschamps & Jörg Matschullat
October 2010

Obituary for Wolfgang Höll (†)

Apart from normal obstacles and challenges that every larger research project faces, the ARSENEX project suffered from a substantial and very sad loss. Early in 2010, a very good friend, a long-time advisor, role model and moral support, a dedicated scientific mind, and a warm-hearted personality with a genuine joie-de-vivre, Prof. Dr. Wolfgang Höll (*1944, †2010) died. He could no longer fight the disease that had taken hold of his body. Wolfgang was Deputy Head of the Institute for Technical Chemistry, Section WGT, at the Karlsruhe Research Centre (today KIT), and a world-renowned researcher.

His research profile in water technology and treatment can best be described by the keywords "selective sorption, ion exchange, magnetic micro ion exchangers, magnetic separation processes, biosorbents, soil/groundwater remediation, mathematical modelling and performance prediction". Wolfgang was strongly linked to many international partners, among them Austria, Brazil, Canada, China, New Zealand, Romania, Taiwan, and Turkey. Brazil was one of his beloved regions and he invested considerable time in learning Portuguese to further improve his generally outstanding communication abilities. Such dedication won him many hearts and he always rejoiced in being able to talk to the children, the shop-keepers and restaurant staff, as well as anybody else who crossed his path during field work and his research stays in the country.

We, the editors and the entire project team, miss him. He remains in our hearts and his modesty and expertise will keep guiding us.

Belo Horizonte and Freiberg, October 2010, Eleonora Deschamps and Jörg Matschullat.

Figure 0.2. Wolfgang in his element, here the water purification plant in Bad Rappenau, Germany, with his innovation, the CARIX process (photo 2001 courtesy of Chiung-Fen Chang, Taiwan, and Ursula Dumas-Höll, Germany).

About the authors

The authors present themselves in alphabetical order (family names).

KATIANE CRISTINA DE BRITO ALMEIDA

Degree in Biology, a specialist in environmental management of water resources. Former consultant of the Water Management Institute of Minas Gerais (Instituto Mineiro de Gestão das Águas), where she conducted a series of studies, measurements, and processing of data on qualitative monitoring network of water resources. Currently, she participates in research projects in the das Velhas river basin and coordinates the integrated monitoring group of the das Velhas basin (Monitoramento Integrado da bacia do Rio das Velhas).

NEILA ASSUNÇÃO

Environmental Engineer, doing her post-graduation in workplace safety engineering at FUMEC, Belo Horizonte. Experience in management systems and environmental permits. She is currently working as an environmental consultant at FEAM (Fundação Estadual do Meio Ambiente), the State Environmental Agency of Minas Gerais in Belo Horizonte.

EDUARDO MELLO DE CAPITANI

Physician with a Masters Degree and Doctorate (PhD) in collective health from the State University of Campinas (UNICAMP), São Paulo, Brazil. Specialist in labour medicine and public health. Professor at the College of Medical Sciences of UNICAMP. Vice-coordinator of the Intoxication Control Center.

NILTON DE OLIVEIRA COUTO E SILVA

Chemist with a B.Sc. Degree from the Federal University of Viçosa (UFV). Masters Degree in nuclear technical sciences from the Federal University of Minas Gerais (UFMG). Currently, he is a health and technology analyst at the State Health Authority FUNED (Fundação Ezequiel Dias) in Belo Horizonte. Extensive experience in chemistry with emphasis on analytical instrumentation (inductively coupled plasma optical emission spectrometry, ICP-OES).

ELEONORA DESCHAMPS

Chemical Engineer (B.Sc.) from the Federal University of Minas Gerais (UFMG) in Belo Horizonte. Engineering Degree (Dipl.-Ing./M.Sc.) from the Rheinisch-Westfälische Technische Hochschule (RWTH) in Aachen, Germany. Doctoral Degree (PhD) in metallurgic engineering from the College of Engineering at UFMG, sandwich course with TU Bergakademie Freiberg, Germany. Today, Eleonora is a professor at the FUMEC University Center of Environmental Engineering, and works as manager of the Division for Industrial Solid Waste at the FEAM Department for Industrial processes and Environmental Analysis, Belo Horizonte. Competence areas: ore treatment, environmental geochemistry, controlled waste disposal, licencing and control of mining activities and chemical industry.

LEONARDO FITTIPALDI

Biologist. Masters in sanitation, environment and water resources. Representative of FEAM in the environmental education area of FEAM in the Interinstitutional Commission of Minas

Gerais. Member of the Environmental Adaptation Program of FEAM for environmental education in public buildings of Minas Gerais – environmental education in the ARSENEX project, IGAM/FEAM. Currently, he works as a technician in the Extension Department of Environmental Education, FEAM in Belo Horizonte.

WOLFGANG HÖLL († 2010)

Chemical engineer. Professor at the Karlsruhe Institute of Technology (KIT), in Karlsruhe, Germany. Senior Scientist. Chairperson of the Division of Water Technology, Deputy Director of the Water Chemistry Institute, Section WGT, of the former Federal Research Centre Karlsruhe (FZK), now part of KIT. Relevant competence areas: ionic change, adsorption, soil remediation. Visiting professor and researcher in various international cooperation projects.

JÖRG MATSCHULLAT

Geologist (B.Sc.), graduated from the Technical University of Clausthal, Masters Degree (Dipl. Geol.) from the University of Tübingen, Dr. rer. nat. in geology and geochemistry from the University of Göttingen. Post-Doctorates at the universities of Clausthal and Heidelberg, Germany. As of 1999, Jörg is full professor of geochemistry and geoecology at the Technical University Bergakademie Freiberg, Director of its Interdisciplinary Environmental Research Centre (IÖZ), and Dean of the Faculty of Geosciences, Geoengineering and Mining of TU Bergakademie Freiberg. Areas of interest: atmospheric and climate science, environmental geochemistry and Earth system science.

SILVÂNIA MATTOS

Pharmacist-biochemist, Master's Degree and Doctorate (PhD) in chemistry from the Federal University of Minas Gerais (UFMG) in Belo Horizonte. Former researcher at FUNED (Fundação Ezequiel Dias), where she coordinated research projects on metals in exposed populations, food and water. Currently, she is a researcher at ANVISA, Brasília, on laboratory quality management.

JAIME MELLO

Agronomist from the Federal University of Santa Maria. Masters Degree in biodynamics and soil productivity from the Federal University of Santa Maria (UFSM) in Rio Grande do Sul. Doctorate in soil and plant nutrition from the Federal University of Viçosa (UFV) in Viçosa, Minas Gerais. Post-Doctorate from the Universidad de Cordoba (Spain) and the University of Illinois in Urban-Champaign (USA). Associate professor of soil chemistry of the Department of Soils at the Federal University of Viçosa. He has extensive experience in agronomy, with emphasis on soil chemistry, developing mainly the following themes: soil geochemistry and flooded sediments, cycling of residues in soils, heavy metals, arsenic and acid drainage.

ISABEL MENESES

Geographer with a Masters Degree from the Catholic University of Minas Gerais (PUC Minas) in Belo Horizonte. Currently, she is an environmental analyst for FEAM (Fundação Estadual do Meio Ambiente) and professor at the Environmental Administration Course of Faculdades Arnaldo Janssen at Belo Horizonte, Minas Gerais. She has many years of experience in geosciences, with emphasis on geoecology, developing the following themes: evaluation of environmental impacts, environmental monitoring, geoprocessing.

SANDRA OBERDÁ

Bachelors Degree in Chemistry and Masters Degree in analytical chemistry from the Federal University of Minas Gerais (UFMG) in Belo Horizonte. Today, Sandra works as an environmental consultant with experience in chemical and environmental analyses, and as a researcher in projects for CDTN and FEAM. She has experience in environmental permits and environmental monitoring.

HELENA EUGÊNIA LEONHART PALMIERI

B.Sc. in Chemistry and Masters Degree in sciences and nuclear techniques from the Federal University of Minas Gerais (UFMG) in Belo Horizonte. Doctoral Degree (PhD) in natural sciences, environmental geology and conservation of natural resources from the Department of Geology of the Federal University of Ouro Preto. Researcher at CDTN-CNEN, Belo Horizonte, working in analytical chemistry, trace analysis and environmental chemistry.

FRIEDRICH EWALD RENGER

Geologist, graduated from the Free University Berlin, Dr. rer. nat. from the University of Heidelberg (1969) – with a thesis on the southern Serra do Espinhaço, Minas Gerais, Brazil. Director of the Eschwege Institute in Diamantina, MG (1970/74), now part of the Institute of Geosciences of UFMG; From 1974 to 1993 exploration geologist in Brazil and Bolivia. Since 1993 professor at the Institute of Geosciences of UFMG; retired since 2008. Areas of interest: regional geology of the Serra do Espinhaço and Iron Quadrangle, geoconservation and geological heritage, history of mining and of geological sciences in Brazil.

ADRIANO TOSTES

Sociologist with degrees (B.A., M.A.) from the Federal University of Minas Gerais (UFMG) in Belo Horizonte. Currently, Adriano works as an analyst for FEAM (Fundação Estadual do Meio Ambiente). He has extensive experience in sociology with emphasis on the fundaments of sociology.

OLÍVIA MARIA VASCONCELOS

Chemical engineer with a degree from the Federal University of Minas Gerais (UFMG), and currently works as a technical coordinator of the Laboratories of Sanitation and Environmental Engineering at UFMG in Belo Horizonte. Her specialty lies in analytical chemistry (atomic absorption spectometry and gaseous and liquid chromatography), in the detection of environmental contamination (trace metals, cyanates, VOC, PAH, pesticides, and others) and in the speciation of sulphur compounds.

RAQUEL VIEIRA

Lawyer with a degree from Faculdade Milton Campos, certified by OAB. She is a specialist in environmental management with a post-graduate degree in business and entrepreneurial law. She worked as a consultant for FEAM (Fundação Estadual do Meio Ambiente) in Belo Horizonte. She is also member of the team of Carneiro & Souza Advogados Associados, a law firm specializing in environmental law. Currently, she is working as an environmental juridical consultant for the Industry, Commerce and Services Subsecretary (SICS) of the State Economic Development Secretariat.

ZENILDE DAS GRAÇAS GUIMARÃES VIOLA

Chemist with a B.Sc. Degree and a Masters in sanitation, environment and water resources from the Federal University of Minas Gerais (UFMG). Zenilde is currently pursuing her doctorate at the same university, and developing a research project for Fundação Centro Tecnológico of Minas Gerais. She works in the Water Management Institute of Minas Gerais. Zenilde has extensive experience in chemistry with emphasis on environmental chemistry, working mainly on water quality, analytical chemistry and ecology.

Section I
Arsenic in the environment, toxicology and remediation

CHAPTER 1

The global arsenic cycle revisited

Jörg Matschullat

1.1 MOTIVATION AND CHALLENGES

There is little doubt about the toxic qualities of arsenic (As), and in particular, the toxic effect of some As species on biota, including humans (▶ 2). True contamination and pollution are usually highly restricted to location, however, and mostly relate to natural and anthropogenic point sources. Recent geochemical mapping projects in Europe and Japan with both highly industrialized and hydrothermally active areas demonstrate this clearly (Imai *et al.* 2004; Reimann *et al.* 2009; Salminen *et al.* 2005). As scientists, we shall contribute to a prudent and objective discussion to explain phenomena and processes and to support rational action and political decisions needed where this element truly gives reason for concern.

Ten years ago, the author published a review paper on arsenic in the environment (Matschullat 2000). Around that time, others contributed to a better understanding of arsenic in the environment, e.g., Buat-Ménard *et al.* (1987), Chilvers and Peterson (1984), Cullen and Reimer (1989), Fowler (1983), Han *et al.* (2003), Hutchinson and Meema (1987), Mandal and Suzuki (2002), Nriagu (1994), Plant *et al.* (2005), Reimann *et al.* (2009), Savory and Wills (1984), Smedley and Kinniburgh (2002), Stoeppler (2004), Tamaki and Frankenberger (1992). Quite some understanding has grown ever since. To comprehend the recent progress in this topic, it is useful to succinctly summarize the current discussion and deliver an update. The subsequent chapters of this book may synthesize recent findings to further a coherent discussion about As-related issues.

Historically, arsenic played an ambivalent role for humans due to its application for healing purposes, murder and suicide (Cullen 2008). The element was known as early as the Bronze age as an admixture component to brighten the colour of bronze (Riederer 1987). Trueb (1996) mentioned bronze items from the 3rd Millenium BC with As contents of 4–5%. The early Chinese, Indian, Greek and Egyptian civilizations mined As-bearing minerals already (Azcue and Nriagu 1994). There is no doubt that even the toxic qualities were discovered early on. Written evidence from China (222 BC) describes pharmaceutical products made from realgar (AsS; Lin *et al.* 1998). Presumably, Albertus Magnus synthesized the element in 1250 AD by heating soap with orpiment (As_2S_3; Schröter *et al.* 1983; Winter 2010). Paris Green, also called Schweinfurt Green, [$Cu(C_2H_3O_2)_2 \cdot 3Cu(AsO_2)_2$)], gained widespread use in paint for artists, wallpaper production and pesticides in the 19th Century. The compound is highly toxic and responsible for numerous cases of arsenicosis (Andreas 1996). The As-based drug Salvarsan (Arsphenamine) was synthesized in the early 20th Century and applied as a common medication to treat syphilis (▶ 2). This supposed familiarity with the element might be one reason for relative carelessness in peoples' behaviour that can be observed in many countries even today (somewhat comparable to the mercury story; Lindqvist *et al.* 1991).

Arsenic once again draws public attention since news about epidemic-like health problems of many people in Bangladesh and West Bengal were reported around the world. There, problems triggered by elevated As values in groundwater remain an issue (Abernathy *et al.* 1997; Chatterjee *et al.* 1995; Das *et al.* 1995, 1996; SEGH 1998). An increasing number of As-related issues became public as of the 1990s, accompanied by an ongoing series of multidisciplinary conferences (Bhattacharya *et al.* 2007; Bundschuh *et al.* 2005, 2006, 2009; Chappell *et al.* 1999, 2001, 2003), dating back to a meeting of the Society of Environmental

Geochemistry and Health (SEGH) in 1991. These meetings have shown that As-related problems occur in very many countries. Arsenic release can be traced back to several major sources, namely volcanism, hydrothermal activity, gold and base metal mining and processing, and agricultural and industrial practices. In Latin America, As pollution is being reported from most countries, e.g., Argentina, Bolivia, Brazil, Chile, Cuba, Ecuador, El Salvador, Mexico, Nicaragua, Peru, Uruguay (Bundschuh *et al.* 2008, 2009).

The attention is primarily driven by the potential detrimental health effects of elevated As exposure (▶ 2; Das *et al.* 1995; Smith *et al.* 1992). Recently, new insights emerge about volatile As species that may be more common in the environment than previously assumed (Planer-Friedrich *et al.* 2006) and that may pose significant health risks despite their very limited occurrence.

This chapter provides an overview of the role of arsenic in the bio-geosphere, discusses element sources and sinks and presents an updated global budget. Individual examples illustrate the current state of knowledge. It becomes clear that even now our understanding of the global As cycle is limited. Significant knowledge gaps exist for all environmental compartments – air, water, sediments, soils, and biota – and limit a reliable quantification especially of natural As sources (Buat-Ménard *et al.* 1987; Matschullat 2000; Fig. 1.1).

1.2   THE GLOBAL BIOGEOCHEMICAL ARSENIC CYCLE

Despite plenty of data on As concentrations and fluxes in the different environmental compartments, many published quantitative data are questionable or outdated. Next to analytical

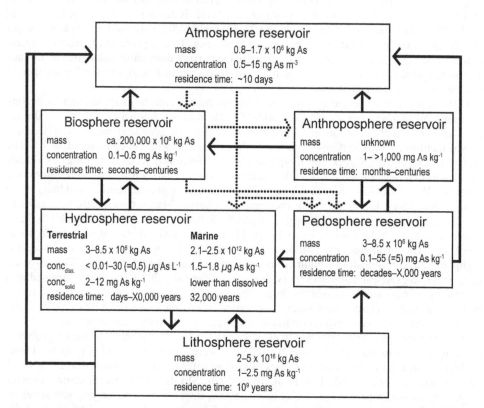

Figure 1.1.   The global As cycle. This figure uses data from the figures 1.2–1.5 (rounded values for clarity). See text for more detail.

problems and related errors (Ebdon *et al.* 2001; Feldmann *et al.* 2004, Plant *et al.* 2005), there is a potential for severe methodological faults related to sampling; particularly critical for water samples (McCleskey *et al.* 2004; Smedley and Kinniburgh 2002; Stoeppler 2004). Related data have to be scrutinized with great care. Therefore, not only do data collected prior to the 1980s need to be questioned – these were often based on limited analytical resources – but also many of the currently published As data should be thoroughly scrutinized prior to use. Many of the data in this book may be representative for the described region only (▶ 6, 7). This chapter, however, attempts to compile data of the highest possible reliability and representativity. It has to be pointed out that any endeavour to make global assessments of element balances and budgets, requires not only a robust database of element concentrations and fluxes for the individual compartments, but also precise information about the related reservoirs. Even this basic requirement often cannot be met with desirable accuracy and only plausible approximations appear reasonable. This illustrates that many questions remain although an element like arsenic has been known and studied for such a long time.

Lithosphere. Arsenic occurs in the environment in the oxidation states –III, 0, III, and V (Stoeppler 2004). The element is mainly associated with sulphide minerals (chalcophilic). The most important As-bearing minerals are mixed sulphides of the $M^{(II)}AsS$ type, where $M^{(II)}$ stands for Fe, Ni, and Co and other bivalent metals (arsenopyrite FeAsS, realgar AsS, niccolite NiAs, cobaltite CoAsS) – (see Allard 1995, Onishi 1969, Plant *et al.* 2005, and Reimann and de Caritat 1998, for a much more complete list of the known As minerals). Arsenates, e.g., mimetesite, $Pb_5[Cl(AsO_4)_3]$, reveal close crystal-chemical relations to phosphates and vanadates. Pentavalent arsenic (As^V) shows strong similarities with phosphorous in phosphates due to similar ionic radii (Goldschmidt 1958).

The undifferentiated Earth's crust contains As concentrations between 1 and 2.5 mg kg^{-1}, based on different calculations of concentrations in rock material (Table 1.1). With a crustal mass of 2×10^{22} kg* (Wedepohl 2004), the total As amount is estimated to $2–5.0 \times 10^{16}$ kg (Reimann and de Caritat 1998: 4.01×10^{16} kg). Natural As liberation from the lithosphere into the exogenic cycle (17.2×10^6 kg a^{-1}) is dominated by terrestrial volcanic exhalations and eruptions. Another 4.9×10^6 kg a^{-1} have to be added, estimated to stem from submarine volcanism (Figs. 1.1, 1.2) that will rapidly become part of the oceanic hydrosphere. However, emission data from volcanic activity may easily vary by approximately two orders of magnitude (Chilvers and Peterson 1987), and the given data may rather underestimate longer-term reality.

Mining and related processes deliver 20.26×10^6 kg of arsenic (in 2009; Brooks 2010) – making these processes equally strong as volcanism. Compared to previous decades, the total anthropogenic As consumption has decreased considerably (Stoeppler 2004). Lithosphere input may be calculated with crustal subduction that yields and transports part of the annual marine sedimentation ($46.4–104 \times 10^6$ kg As a^{-1}). Subduction has been estimated as mounting to 38.2×10^6 kg As a^{-1} in equilibrium with production at mid-ocean ridges (Wedepohl, 2004; this work; subduction rate: ø 5.5 cm a^{-1}; subducted share of oceanic crust: 12.7 km^3 a^{-1} – after data from Toksoz 1976; assumed density: 3 g cm^{-3}). Thus, input into the lithosphere (based on assumed subduction rates) roughly equals the output and appears balanced (Fig. 1.2).

Atmosphere. Arsenic dominantly occurs in the species arsine, metallic arsenic, inorganic trivalent and pentavalent arsenic, organic monomethylarsenic acid (MMA), dimethylarsenic acid (DMA) and/or their salts in the atmosphere (Spini *et al.* 1994). More, often shortlived, but highly toxic organic species may occur close to the surface (Planer-Friedrich *et al.* 2006). An average of 1.74×10^6 kg As is stored in the troposphere (higher atmospheric layers can be neglected in this mass balance), albeit unevenly distributed between the hemispheres with 1.48×10^6 kg in the northern and 0.26×10^6 kg in the southern hemisphere (Chilvers and

---

*Other sources: upper crust 1.13×10^{22} kg, total crust 2.13×10^{22} kg (Wedepohl 1995), lithosphere 1.365×10^{22} kg (Tye Morancy – www.madsci.org)

Table 1.1. Arsenic concentrations (mg kg^{-1}) in the lithosphere and in extraterrestrial objects, given in full ranges and in brackets as more likely global average concentrations ranges.[#]

Medium	Concentration	Source
Universe	0.008	19
Stony meteorites	1.8–18	11, 19
Iron meteorites	11	11
Earth crust, total[*]	1.0–2.5	4, 15, 16, 17
Continental crust[*]	1.0–4.8	14, 15, 16, 17, 18
Oceanic crust[*], MORB	1.0–1.5	3, 18
Ultramafic rocks	0.3–15.8 (0.7)[#]	3, 5
Gabbroic rocks, basalts	0.06–113 (0.7–1.5)[#]	3, 5, 18
Granitic rocks	0.18–15 (1.5–3.0)[#]	3, 5, 18
Rhyolites	3.2–5.4	5
(grano)diorites, syenites	0.09–13.4	5
Metamorphic rocks	0.5–11	11
Gneisses, mica schists	<0.1–18.5 (4.3)[#]	5, 18
Slates, phyllites	0.5–143	5
Quartzites	2.2–7.6	5
Granulites	1.3	18
Shales; schists	3–490 (10–13)	3, 5, 11, 18
Sedimentary rocks	0.1–188	1, 3, 5, 11, 18
Sandstones	0.5–9 (1.0)[#]	3, 5, 11
Greywackes	8	18
Carbonates, limestones	0.1–20.1 (1.0–2.5)[#]	1, 3, 5, 11, 18
Phosphates	0.4–188 (12)[#]	5, 11
Coal	0.15–35,000 (10)[#]	2, 10, 13, 14[#]
Brown coal (lignite)	0.34–130; (5–45)[#]	12
Hard coal	21	18
Crude oil	0.0024–1.63; (0.13)[#]	12, 18

[*]Mass Earth crust: 2.13×10^{22} kg; Mass upper crust: 1.13×10^{22} kg after Wedepohl (1995, 2004); MORB: mid-ocean ridge basalt; [#]data in brackets appear most representative for the respective matrix. 1 Chester (2000), 2 Finkelman *et al.* (1999), 3 Koljonen (1992), 4 Lide (1996), 5 Mandal and Suzuki (2002), 10 Mukherjee *et al.* (2008), 11 Onishi (1969), 12 Pacyna (1987), 13 Plant *et al.* (2005), 14 Reimann and de Caritat (1998), 15 Rudnick and Gao (2004), 16 Taylor and McLennan (1995), 17 Wedepohl (1995), 18 Wedepohl (2004), 19 Winter (2010).

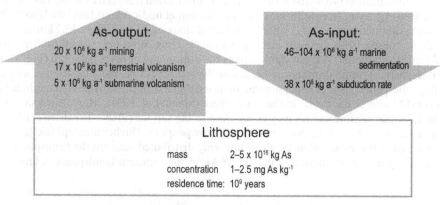

Figure 1.2. The lithospheric As reservoir, related fluxes and residence time (see text for details).

Table 1.2.   Hemispheric and continent – landmass distribution of atmospheric deposition.

	Northern hemisphere	Southern hemisphere	Sources
On landmasses	20×10^6 kg As a^{-1}	7.6×10^6 kg a^{-1}	Duce pers. comm. 1999
Over oceans	2.2×10^6 kg a^{-1}	0.6×10^6 kg a^{-1}	
Total	73.6×10^6 kg a^{-1}	4.3×10^6 kg a^{-1}	Chilvers and Peterson (1987)

Peterson 1987; Figs. 1.1, 1.2, Table 1.2). This imparity is explained by the larger land mass and the abundance of highly industrialised countries with their particular emission profiles on the northern hemisphere. Nevertheless, the data should be questioned due to substantial natural As emissions. Walsh *et al.* (1979) calculated a distinctively lower atmospheric As mass (0.8×10^6 kg), however, based on a slightly thinner tropospheric layer and different numbers for anthropogenic emissions. Chilvers and Peterson (1987) included the latest emission data as well as the recent development, e.g., of the emerging Asian industrial countries. This may be outdated by now, because improved environmental technologies have reduced related atmospheric emissions in many parts of the world. With 2.8 ng m^3, their As concentration value for the northern hemisphere may still be too high if the following considerations are accepted.

Within the troposphere, particulate As dominates with 89–98.6% as compared to gaseous arsenic. Thus, atmospheric As retention time is between seven (Rahn 1976) and almost 10 days (Duce pers. comm. 1999). Dusts are the major form or transport medium for anthropologically induced metallic As emissions (sulphides and oxides; ▶ 10). The average effective diameter of As-containing aerosols from combustion processes is ~1 μm. Clearly, the emission profile shows a relationship with the applied scrubber or cleaning techniques (Pacyna 1987; Table 1.6). The total annual As input into the atmosphere (natural and anthropogenic) was calculated as being approximately between 73.54×10^6 kg a^{-1} (Chilvers and Peterson 1987) and 296.47×10^6 kg a^{-1} (Mackenzie *et al.* 1979), the latter based upon a high estimate of natural As release via marine aerosols and biogenic liberation. It seems plausible to assume that more recent data are more realistic, because these include more direct measurements of natural releases. Biogenic liberation includes low temperature release via metabolic activities in soils, a natural source, whose influence is being addressed rather differently. Chilvers and Peterson (1987) quantified it with $0.16–26.2 \times 10^6$ kg a^{-1}. The higher number appears more realistic due to previous underestimates for soil microbial metabolism. Sixty percent of the anthropogenic As emissions account for two sources: Cu-smelting and combustion of fossil fuels. Other sources include the application of herbicides, Pb and Zn smelting, glass production, wood preservation, waste incineration and steel-production, deforestation of tropical woods, savannah and temperate woodlands, burning of grassland, use of wood as fuel. These anthropogenic emissions total 28.07×10^6 kg a^{-1} (Chilvers and Peterson 1987; see anthroposphere; Tables 1.3, 1.6; Figs. 1.1, 1.2).

Deposition from the troposphere is highly differentiated between both landmasses and oceans, and between the northern and the southern hemisphere (Table 1.2). For the oceans, Duce *et al.* (1991) gave $2–5 \times 10^6$ kg As a^{-1} as dissolved input and $1.3–3.0 \times 10^6$ kg a^{-1} as particulate As input, as compared to a total mass of 4.3×10^6 kg a^{-1} (Chilvers and Peterson 1987). Chester (2000) compiled atmospheric fluxes onto the ocean surface with 3–280 ng As cm^{-2} a^{-1}. Assuming similar fluxes onto terrestrial surfaces, a range of $15–1,400 \times 10^6$ kg a^{-1} results, making the higher numbers rather very unrealistic. Due to the limited residence time, masses must be roughly in balance with global emissions of 73×10^6 kg a^{-1} and inputs (deposition) of $30–80 \times 10^6$ kg a^{-1} (Fig. 1.3). Since those balances are partly self-referential, it becomes obvious that the natural As fluxes need better quantification.

Table 1.3.   Global atmospheric As fluxes and As concentrations[a].

Emission	As flux (10^6 kg a^{-1})[b]	Deposition	As flux (10^6 kg a^{-1})[b]
Cu-smelting	12.08	N hemisphere	22–74
Coal combustion	6.24	S hemisphere	4.3–8.2
Herbicide use	3.44	**Sum deposition**	**26.3–82.2**
Pb and Zn-smelting	2.21		
Tropical deforestation	1.60	Remote areas[c]	0.0009–80 ng L^{-1}
Grassland flaming	1.00	Rural areas[c]	240–370 ng L^{-1}
Glass production	0.467	Urban areas[c]	0.5–16 µg L^{-1}
Wood as fuel	0.425		
Other deforestation	0.320	Remote areas[d]	0.008–1 ng m^{-3}
Wood preservation	0.150	Contaminated areas[d]	15 ng m^{-3}
Waste incineration	0.078		
Steel production	0.060	Remote areas[e]	0.31 g ha^{-1} a^{-1}
Low T soil release	0.16–26.2	Rural areas[e]	3–10 g ha^{-1} a^{-1}
Other natural	~19	Urban areas[e]	300 g ha^{-1} a^{-1}
Sum emissions	**28.23–54.27**		

[a]Mass of the atmosphere: 5.1×10^{18} kg (Jacob 1999; Williams 2010); worldwide precipitation: 0.505×10^{18} kg a^{-1} (Berner and Berner 1996); [b]data after Chilvers and Peterson (1987) and Duce (pers. comm. 1999); [c]concentration bulk deposition; [d]concentration in air; [e]deposition; data compiled from Reimann and de Caritat (1998), Nriagu (pers. comm. 1999) and Plant *et al.* (2005).

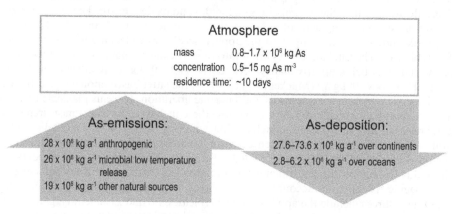

Figure 1.3.   The atmospheric As reservoir, related fluxes and residence time (see text).

Using some basic data, the scenarios can be tested independently. Given a global precipitation average of 0.505×10^{18} L a^{-1} and a concentration therein of 80 ng As L^{-1} (Table 1.3), an annual atmospheric deposition of 40.4×10^6 kg As results; obviously somewhat too low. Using Andreae's data (1980), the background concentration would have to be calculated with 4–19 ng As L^{-1} in precipitation only. A total planetary surface of 5.1×10^{10} ha and an atmospheric As deposition of 0.3 g ha^{-1} a^{-1} (Table 1.3) result in 15.3×10^6 kg As a^{-1}. Both calculations use As data for clean air regions (remote areas). Adding, e.g., 50% of this number to account for the anthropogenic As release, the fluxes would be somewhat balanced. These simple examples illustrate the necessity of further work to increase the reliability of atmospheric As budgeting.

Reimann and de Caritat (1998) give bulk deposition values of 80 ng L^{-1} (remote areas), 370 ng L^{-1} (rural areas), and 12.3 µg L^{-1} (polluted areas). The same authors compiled As concentrations in air of 0.5–1 ng m^{-3} (remote) and 15 ng m^{-3} (polluted areas). Cloud water in Central Europe (Brocken, Harz) yielded a median value of 0.24 ng As m^{-3}, range 0.04–4.4 ng m^{-3} (0.4 µg L^{-1}, range 0.1–6.6 µg L^{-1}), according to Plessow *et al.* (2001) – in very good agreement with the data from Reimann and de Caritat (1998). Nriagu (pers. comm. 1999) suggested As concentrations in wet deposition of 10 ng L^{-1} (remote), 240 ng L^{-1} (rural) and 900 ng L^{-1} (urban areas). Andreae (1980) published As concentrations of 19 ng L^{-1} in marine rain (Hawaii and Californian coast). Fergusson (1990: 177) compiled recent data on concentrations in remote area air with 8–160 pg m^{-3} and in precipitation with 0.9–37 ng L^{-1}, as well as atmospheric deposition with 0.31 g ha^{-1} a^{-1}. Plant *et al.* (2005) compiled precipitation data mainly from the United States with concentrations between <0.005 and 1.1 µg L^{-1}, stating that non-polluted precipitation contains less than 0.03 µg L^{-1}.

Latest data (2008–2010) from the eastern Erzgebirge (Germany) show remote precipitation concentrations <0.2–0.9 µg L^{-1} (median <0.2 µg L^{-1} for open field deposition; the high values occurred exclusively in the heating period (local lignite; Matschullat *et al.* unpubl.). In the early 1990s medians were between 0.7 and 3.2 µg L^{-1} in open and throughfall deposition, respectively – more representative then for Central Europe with still abundant As sources. Those older data correspond to a deposition of 6.3 and 13.5 g ha^{-1} a^{-1} (Erzgebirge), and 7.8 and 31.4 g ha^{-1} a^{-1} (Harz Mountains), while the new ones give 0.9 g ha^{-1} a^{-1} (eastern Erzgebirge). Aerosol concentration in the early 1990s was on average between 3.2 and 3.6 ng m^{-3} (eastern Erzgebirge), with a distinct dominance during the winter heating period (Matschullat and Kritzer 1997). A strong decline of As deposition via precipitation from 11 g ha^{-1} in 1984, 3.0 g ha^{-1} in 1993 to <1 g ha^{-1} in 2009 was measured in Central Europe and in Germany (Matschullat *et al.* unpubl.). This decrease runs parallel to the As emission reductions from approximately 82 × 10^3 kg in 1985 to 21 × 10^3 kg in 1995 (only western Germany; Schulte and Blum 1997). The deposition decline may also be inferred from results of the ECE-moss-monitoring programme: minima on Iceland (0.06 mg kg^{-1}) and maxima in Romania (0.95 mg kg^{-1}) represent the range of national median values in all of Europe. A comparison of related moss mean values from Scandinavian countries from 1975 (0.75 mg kg^{-1}) and 1990–1995 (0.23 mg kg^{-1}) to 2000 (0.12 mg kg^{-1}) substantiates this observation (Kabata-Pendias and Mukherjee 2007) – and indicates a decline of roughly 80%. (Within Germany, elevated values were measured mainly in mosses from Saxony, close to the Czech border, maxima 1.2 mg kg^{-1}). In total, As concentrations in moss from Germany ranged in 1995 from 0.001 to 2.7 mg kg^{-1} (median 0.25 mg kg^{-1}; Siewers and Herpin 1998). Compared with data from 1990 to 1991 (median 0.34 mg kg^{-1}) this represents a decrease by approximately 24% (Rühling 1994; Rühling and Steinnes 1998). In epilithic lichens (*Lecanora muralis* from Germany) As concentrations between 0.14 and 166 mg kg^{-1} (median 3.4 mg kg^{-1}) were determined. Elevated values were always bound to plausible local anthropogenic emission sources (Matschullat *et al.* 1999). In the same paper, data from the crustose lichen *Xanthoria elegans* (Ontario, Canada) show a range of 0.9–24 mg As kg^{-1} (median 1.9 mg kg^{-1}). Again, a direct source – receptor response was shown for examples with elevated As concentrations.

Special conditions, characterised by urban microclimate and by particular emission fingerprints, prevail in urban and highly industrialised areas, and are reflected by urban aerosols and dusts. In urban areas, approximately 20% of the emitted arsenic stems from high temperature combustion (anthracite burning > waste incineration > lignite burning > cement works), and 55% from road traffic (gasoline soot > diesel soot > brake wear > tire wear; Heinrichs and Brumsack 1997). Less than 2% of the As input are from natural sources. Compared with a non-polluted situation, atmospheric enrichment factors sensu Lantzy and Mackenzie (1979) can be calculated and show Al-normalised EF of 10^2–10^3 in urban dusts (Heinrichs and Brumsack 1997). It should be noted, however, that the use of such enrichment factors yield some risk of major misinterpretation (Reimann and de Caritat 2000, 2005). For human susceptibility, however, indoor air pollution at least within urban areas deserves as much attention as outdoor atmospheric deposition (▶ 10). Within apartment walls, mildew may biomethylate arsenic into trimethylarsine ($(CH_3)_3As$), a foul smelling and highly toxic gas (Harrison and Mora 1996) and thus, increase As toxicity, e.g., of As-containing dust (▶ 17).

Pedosphere. Under aerobic conditions, arsenic occurs in soils mainly as arsenate, bound to clay minerals, Fe and Mn-oxyhydroxides, and to organic substances. The bonding strength directly relates to As concentrations, the concentrations of the bonding partners, and the retention time in the related transport process. In acidic soils, Al and Fe-arsenates occur ($AlAsO_4$, $FeAsO_4$), while Ca-arsenate, $Ca_3(AsO_4)_2$, is the dominant species in basic and limy soils (Fergusson 1990). Yet, the chemical forms of arsenates in soil are largely a matter of conjecture. Bowen (1979) compiled As retention times of 1,000–3,000 years for soils under moderate climates. In most tropical soils, these periods may be considerably shorter.

Soils cannot be seen as a homogenous medium and contain highly variable As concentrations (Reimann *et al.* 2009). The humus layer acts as a natural biogeochemical barrier for arsenic and several other trace elements. It suppresses As percolation with seepage water, and thus, strongly accumulates the element (Goldschmidt 1937; Sadiq 1997). Accordingly, the highest natural As concentrations potentially occur in C_{org}-rich soils, explaining very low As concentrations in old tropical soils. Despite these variances, Goldschmidt (1958) calculated average soil As concentrations with 7.5 mg kg^{-1}, based on a soil survey in the USA (range: 4.5–13 mg kg^{-1}). Allard (1995) quoted National Research Council of Canada data from 1978 and suggested a range of 0.1–55 mg kg^{-1} with an average value of 7.2 mg kg^{-1}. Yet, the estimate of 5 mg kg^{-1} for a global As average in soils by Koljonen (1992) is probably more realistic (Figs. 1.1, 1.4; Table 1.4).

European data from the FOREGS mapping exercise give a median value of 7.0 mg kg^{-1} for topsoil and 6.0 mg kg^{-1} for subsoil (Salminen *et al.* 2005). A distinct difference arises between the more northern part of the region and the more southern part that was not directly affected by glaciation (Reimann *et al.* 2009). This difference becomes also apparent when looking at agricultural soils in the Baltic Sea catchment (where anthropogenic enrichment might be assumed). There, median values of 1.9–4.0 mg kg^{-1} were shown for topsoil and 2.0–4.0 mg kg^{-1} for subsoil (ranges: aqua regia to XRF; Reimann *et al.* 2003). Above the northern polar circle, median values for the humus layer (1.16 mg kg^{-1}), the B-horizon (1.1 mg kg^{-1}) and the C-horizon (0.5 mg kg^{-1}) further substantiate this difference (Reimann and de Caritat 1998). Extending further east to the Ural mountains, soils from the entire eastern Barents region delivered median concentrations between 0.98 and 3.30 mg kg^{-1} (avg. median of organic layer 1.1 mg kg^{-1}) and 1.11 and 5.59 mg kg^{-1} (avg. median of C-horizon 1.5 mg kg^{-1}; Salminen *et al.* 2004). Interestingly enough, urban dusts collected on roads, and sediments from rain drains in the 1990's in Germany yielded values similar to the European average (Pleßow *et al.* 1997b; Table 1.4)

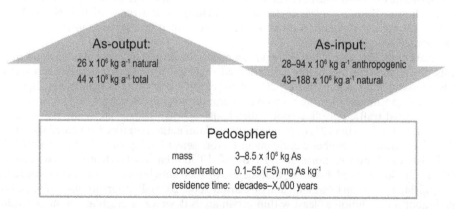

As-output:
26 x 10^6 kg a^{-1} natural
44 x 10^6 kg a^{-1} total

As-input:
28–94 x 10^6 kg a^{-1} anthropogenic
43–188 x 10^6 kg a^{-1} natural

Pedosphere

mass   3–8.5 x 10^6 kg As
concentration   0.1–55 (=5) mg As kg^{-1}
residence time: decades–X,000 years

Figure 1.4.   The pedosphere reservoir, related fluxes and residence time (see text).

Individual areas, however, may show median As concentrations that reach far beyond those natural global averages. Data from Saualpe, Kärnten, Austria (God 1994) and from Feistritz, Lower Austria (God and Heiss 1996) show natural median As values of 100–115 mg kg^{-1} (n = 252) and of 29 mg kg^{-1} (n = 50), respectively. Contaminated soils may yield much higher concentrations that reach into the lower percentage range. Next to atmospheric As deposition via short and long-range transport pathways, As enrichment in soils can be derived mainly from agriculture (pesticides and manure), from mining and smelting activities (e.g., as As_2O_3-emission from smelters; ▶ 12) and as sulphidic As species in waste water and sludge. These emission sources generate local positive anomalies, which may persist as secondary As sources over decades and centuries (Reimann *et al.* 2009). The resulting As enrichment in the upper soil layers may lead to root depression and growth defects in plants and to death of earthworms (Däßler 1986; ▶ 13). In such anomalies, concentrations may easily increase to values far above 100 mg kg^{-1}. Kabata-Pendias and Mukherjee (2007) compiled maximum soil As concentrations between 900 and 2,500 mg kg^{-1} from Asia, Europe and North America; all contaminated from industrial sources.

Matschullat (1996) describes an example of soil contamination by historical mining and smelting activities in the Harz Mountains, Germany. Median concentrations in the soil were 130 mg kg^{-1} (range 7–970 mg kg^{-1}). Even higher concentrations may occur on heaps or in the sediments of tailings ponds, e.g., 300–30,000 mg kg^{-1} (median 10,500 mg kg^{-1}) in tailings from hydrothermal gold processing in Minas Gerais, Brazil (▶ 11; Lottermoser 2010). Fiedler and Rösler (1987) demonstrated effects of As emissions from lignite burning in Germany: As concentrations in spruce needles showed a direct response to distance from the emitter (10–15 km: 8 mg kg^{-1}, 20–22 km: 2.5–4.1 mg kg^{-1}, 30 km: 0.6–0.9 mg As kg^{-1}). Related As accumulation close to the source led to soil As concentrations of 250–1,200 mg kg^{-1} (median: 1,650 mg kg^{-1}). There, As enrichment in vegetables amounted to: lettuce 2–24 (14) mg kg^{-1}, for spinach 11–32 (16) mg kg^{-1}, for parsley 8.5–35 (16) mg kg^{-1}, for carrots 0.3–6 (2.5) mg kg^{-1}, and for grass 3–33 (12) mg kg^{-1} (median values in brackets; ▶ 13). Further As pathways into soils are the different types of surface waste disposal sites, and the illegal disposal of As containing residues. Chilvers and Peterson (1987) estimate the annual input via residues into the pedosphere to amount to 28.4×10^6 kg a^{-1}. Nriagu and Pacyna (1988) calculated the total soil As input with 94×10^6 kg a^{-1} – ca. 41% wastes from commercial products, 23% from coal ashes, 14% atmospheric deposition, 10% tailings materials, 7% smelter residues, 3% agriculture and 2% from industrial production, and further minor sources – probably too high even under the aspects of population growth and material turnover. According to Chilvers and Peterson (1987), the ratio of natural-to-anthropogenic soil As inputs is 60: 40%. This ratio probably has to be re-evaluated following the simple mass balances presented in the section on atmospheric arsenic.

Very few data are available on soil As outputs. The most relevant process seems to be the low temperature volatilisation of organic As compounds from soil and surface sediments into the atmosphere (Chilvers and Peterson 1987; Fowler 1983; see atmosphere). Obviously, an additional export will constantly take place from soil degradation and erosion that deliver into surface and ground waters. As a rough estimate, the particulate load of the streams may serve as an indicator for soil erosion. With 18×10^6 kg As a^{-1}, the total soil export is close to the riverine export number of 23×10^6 kg a^{-1}, Gaillardet *et al.* (2003), and compares quite well with a potential soil input of $28.4–94 \times 10^6$ kg a^{-1} from all other sources (Figs. 1.1, 1.4; Table 1.4).

Hydrosphere. Many As compounds dissolve relatively well in both sea (20–90%, Chester 2000; Neff 2002) and **stream water** (60–80%, Brügmann and Matschullat 1997; 80%, Matschullat *et al.* 1997). Global averages for dissolved arsenic in stream water are typically presented as 0.1–2 μg L^{-1} (<0.02–3.8 μg L^{-1} for non-polluted and up to 45 μg L^{-1} for polluted rivers; Gaillardet *et al.* 2003; Plant *et al.* 2005; Table 1.5) and particulate concentrations as 5 μg L^{-1} (Martin and Whitfield 1983), which appears too high. Mine waters may yield much higher concentrations in the mg L^{-1}-range (e.g., Williams 2001). European data corroborate the lower values and deliver a median value of 0.63 μg L^{-1} (range <0.01–27 μg L^{-1}; Salminen

Table 1.4.　As concentrations, ranges, and pools (mg kg^{-1}) and fluxes (t a^{-1}) in the pedosphere[a].

Parameter	Value	Source
Soils, global	5 (0.1–55) mg kg^{-1}	1, 4, 5
Global soil reservoir[a]	600,000–1,700,000 × 10^6 kg	this work
Retention time, moderate climate	1,000–3,000 years	2
Solid waste from metal fabrication	0.11 mg kg^{-1}	6
Municipal sewage, organic waste	0.25 mg kg^{-1}	6
Fertilisers	0.28 mg kg^{-1}	6
Urban refuse	0.40 mg kg^{-1}	6
Urban street sediment	5.7–7.3 mg kg^{-1}	8
Logging, wood waste	1.7 mg kg^{-1}	6
Animal, agricultural waste	5.8 mg kg^{-1}	6
Atmospheric deposition	13.0 mg kg^{-1}	6
Coal ashes	22.0 mg kg^{-1}	6
Discarded products	38.0 mg kg^{-1}	6
Input, anthropogenic	28.4–94 × 10^6 kg a^{-1}	3, 7
Input, natural	1.5 × anthropogenic	3
Output, natural	26.2 × 10^6 kg a^{-1}	3
Output, total	44 × 10^6 kg a^{-1}	this work

[a]Soil mass: 1:25,000 of the lithospheric mass after Bhumbla and Keefer (1987); 1 Allard (1995), 2 Bowen (1979), 3 Chilvers and Peterson (1987), 4 Goldschmidt (1958), 5 Koljonen (1992), 6 Nriagu (1990), 7 Nriagu and Pacyna (1988), 8 Pleßow et al. (1997b).

et al. 2005) for dissolved arsenic. Respective data for northeastern Europe to the Ural mountains show 0.47 µg L^{-1} (range <0.01–8.75 µg L^{-1}; Salminen et al. 2004) – strongly supporting the calculation for a global average of 0.62 µg As L^{-1} by Gaillardet et al. (2003). The same authors calculate the annual riverine flux into the oceans as 23 × 10^6 kg As a^{-1}.

The As concentration range in **groundwater** is even higher due to possible reducing conditions and longer contact times with the lithosphere. Allard (1995) suggested a background concentration of 0.5–0.9 µg L^{-1}. Driehaus (1994) shows a "normal groundwater range" of 0.01–800 µg L^{-1}, which includes untypical and highly mineralised areas. Plant et al. (2005) provide data between 0.2 and several thousand µg L^{-1}, and define typical values to remain below 10 µg L^{-1}. This value agrees with contamination risks from As-rich groundwater around the world as compiled by Ng et al. (2003). In a semi-active volcanic area of central Italy, Preziosi et al. (2010) determined a natural range of 0.22–128.5 µg L^{-1} (mean 14.5 µg L^{-1}) and show the concentration dependency upon various types of bedrock. Reimann and Birke (2010) recently studied European groundwater using bottled water as a proxy. Their median value of 0.22 µg L^{-1} (range <0.03–90 µg L^{-1}) appears rather realistic for non-contaminated waters, possibly even on a global scale. Pleßow et al. (1997a) identified the distinct increase of As concentrations with decreasing pH-values. If there is available arsenic, its concentrations may exceed the permissible limits for drinking waters (10 µg L^{-1}; ▶ 5) below pH 5. Examples from sandstone aquifers and Paleozoic sediments illustrate the point (Table 1.5). Wood (1974) suggested the release of MMAA and DMAA, built in-situ by anaerobic bacteria within the porewater, into groundwater. This process of organo-As compound formation in anoxic groundwater conditions could not be shown until a few years ago (Bentley and Chasteen 2002).

Increased As concentration may occur in downstream **freshwater sediments** due to changes in pH-values or redox potential, and through water and soil acidification. Matschullat et al. (1994) demonstrated examples from the Harz mountains, Germany (Lake Söse reservoir: 30 mg kg^{-1}) and from the eastern Erzgebirge (Malter reservoir: median 140 mg kg^{-1}; range 19–210 mg kg^{-1}) derived solely from accumulation through acid surges. Catchments with active or derelict mining activities may show similar or even higher enrichments (Borba *et al.* 2003; ▶ 12). Acidification leads to an increase of As mobilisation from affected soils and sediments. Consequently, the resulting As fluxes drain into larger streams. Drastic changes, however, such as an As increase in particulates of the Elbe river at Schnackenburg (Germany) from 20 mg kg^{-1} (1988) to approximately 75 mg kg^{-1} (1994; Müller 1996), are not related to this process. Two extreme high run-off events in 1993 and early 1994 were responsible for the observed phenomenon that mobilised a large fraction of older sediments and, thus, "cleaned" the riverbeds. This is also true for similarly striking changes in metal fluxes of other larger rivers in Germany – and of rivers and streams in tropical and subtropical environments due to highly dynamic and seasonal hydrological variances. This unusual sediment transport has its consequences for the accumulation of trace elements in estuaries and harbour sediments, which have to be dredged at high costs. Calmano (1996) illustrates this point with the 1992 example from Hamburg harbour, Germany: its sediments showed concentrations of 27–65 mg As kg^{-1}.

Terrestrial sediment **pore water** may vary widely (1.3–100,000 µg L^{-1}; Plant *et al.* 2005), with the highest values always pointing at strong pollution sources. In suboxic environments, As concentrations are generally much higher than in the oxic realm. Arsenic is released through reduction of Fe-containing components (▶ 3, 4, 11). In this process, relative As enrichments of up to a factor of 23 were observed, parallel to a mobilisation of Mn, Fe and Cr (Hong *et al.* 1995).

Natural As concentration in stream sediments from large-scale geochemical mapping exercises may yield more representative results. The FOREGS atlas shows a median of 6 mg kg^{-1} for stream and floodplain deposits in Europe (Salminen *et al.* 2005). The nation-wide sediment mapping in Japan delivered a median value of 8 mg kg^{-1} (Imai *et al.* 2004). For comparison, Reimann and de Caritat (1998) quote global sediment concentrations between 2 and 12 mg kg^{-1} from lithologically similar areas that are largely unaffected by acidification processes. In general, average freshwater sediment As concentrations must be expected to be close to average soil As values. Yet, terrestrial geothermal riverine As inputs should be considered, but related data are largely absent.

In **marine waters**, arsenic is present in trivalent, pentavalent, and methylated forms. In O_2-saturated water, arsenate, As$^{(V)}$ (HAsO$_4$), is the dominant dissolved As species, particularly in deep sea water. Respective concentrations of 1.1–1.9 µg kg^{-1} (average 1.7 µg kg^{-1}) at a salinity of 36‰ were given by Bruland (1983). In surface water (photic zone), As$^{(V)}$ is taken up by phytoplankton together with phosphate, and transferred to arsenite (As^{III}), methylarsenate and dimethylarsenate – the methylated species accounting for up to 10% of the total arsenic in the euphotic zone of many oceanic regions. The role of plankton and bacteria cannot be overestimated in As speciation; this is why arsenic is seen as a member of the nutrient-type elements in sea water (Chester 2000). Organo-As compounds of As$^{(III)}$ and As$^{(V)}$ are produced by microbiological processes in sediments and soils, namely monomethylarsenic acid (H$_2$AsO$_3$CH$_3$, MMAA) and dimethylarsinic acid (HAsO$_2$(CH$_3$)$_2$, DMAA), as well as trimethylarsine oxide (AsO-(CH$_3$)$_3$, TMA), phenylarsonic acid (H$_2$AsO$_3$-C$_6$H$_5$, PAA), and some others. Some of this arsenic is liberated into the atmosphere and deposited into the sediments.

Based on an estimated global median value of 1.5–1.8 µg As kg^{-1} sea water (Donat and Bruland 1995), the total As mass in the seas would amount to 2,100,000–2,500,000 × 10^6 kg. These values appear realistic for two reasons. Damm *et al.* (1985a, b) calculated As inputs from hydrothermal activity at oceanic ridges as reaching 4.9 × 10^6 kg a^{-1}, relative to 54 × 10^6 kg a^{-1} from the World rivers. Duce *et al.* (1991) estimated the annual As input through freshwater streams into the oceans as 10 × 10^6 kg a^{-1} dissolved and 80 × 10^6 kg a^{-1} particulate freight. Using a global freshwater mass of 3.6 × 10^{19} kg a^{-1}, transported via streams into the sea (Berner and Berner 1996), an As discharge of 61.2 × 10^6 kg a^{-1} would be obtained with an average dissolved concentration of 1.7 µg L^{-1}. This number appears realistic, but challenges the relatively high estimates for the global particulate input into the seas, because total stream inputs of $183–241 \times 10^6$ kg As a^{-1} cannot be supported by available data. With a sediment flux of $20,000,000 \times 10^6$ kg a^{-1} (Berner and Berner 1996) and a concentration of 10 mg kg^{-1}, the total particulate As transport can realistically be estimated to amount to 20×10^6 kg. Other authors give considerably lower sediment fluxes (e.g., $8,000,000 \times 10^6$ kg a^{-1}, Milliman 1980, or $14,500,000–15,500,000 \times 10^9$ kg Milliman and Meade 1983), but partly higher As concentrations. Particulate concentrations are most likely very low and mostly bound to organic material.

With a global pool of 1.4×10^{21} kg of sea water, and dissolved concentrations of 1.7 µg As L^{-1}, we calculate a total of $2,380,000 \times 10^6$ kg As in the seas. That pool, combined with a residence time of 32,000 years, leads to a flux of 74.375×10^6 kg a^{-1} – a number that reflects a realistic dimension as compared to sedimentation and subduction fluxes and the most likely stream input estimates (Table 1.5; Figs. 1.1, 1.5).

Biosphere. The global biomass is estimated at $2,000 \times 10^{15}$ kg (various non-verifiable internet sources) or at $1,919 \times 10^{12}$ kg (!) of world plant biomass with a share of 184.1×10^6 kg arsenic (from Table 1 in Markert 1992). Table 1.6 presents a compilation of As concentrations in the biosphere. The values give normal concentrations without known additional accumulation, e.g., from contaminated water, soil, fertilizers, or irrigation. The As phytotoxicity decreases in the order arsenite > arsenate > mono-nitrogen-methane-arsonate (MSMA > cacodylic acid, with the latter being highly phytotoxic if directly applied to leaf surfaces; Sachs and Michael 1971). The highest As concentrations occur in plant roots, the lowest generally in fruit and seeds. Only soil As values that are far above average (200–300 mg kg^{-1}) lead to elevated As

Table 1.5.   Arsenic concentrations (µg L^{-1}) and fluxes (kg a^{-1}) in the hydrosphere[a] (see text).

Medium	Concentration/flux	Source
Stream water, dissolved	0.1–1.7	2, 4, 13
Stream water, particulate[b]	5	12
Groundwater, global	0.1–230	1
Groundwater, background	0.5–0.9	1, 13
Groundwater, S-Norway	0.18 (0.011–19)	15
Groundwater, Germany	<0.05–150	13, 14, 16, 18
Sandstone aquifers	2.5 (0.5–11.2)	18
Palaeozoic sediments	0.8 (0.5–25)	18
Acidified groundwater	0.28 (<0.05–35)	14
All aquifer types	<0.5–1	16
Groundwater, West Bengal	190–740	3, 6
Seepage from waste disposal	3–30	10
Brackish water, estuary	<4	13
Sea water, dissolved	1.5–1.7 µg kg^{-1}	7
Sea water, particulate	13 ng kg^{-1}	7
Ocean residence time	$3.2–6.3\ (5) \times 10^4$ years	4, 17
River input into seas	$23–100 \times 10^6$ kg a^{-1} (23*)	4, 8, 9*, 11, 13
Submarine volcanism	4.9×10^6 kg a^{-1}	5

[a]Mass of the hydrosphere 1.459×10^{21} kg (Berner and Berner, 1996), separated into oceans (1.4×10^{21} kg), continental ice caps and glaciers (43.4×10^{18} kg), ground water (15.3×10^{18} kg), lakes (0.125×10^{18} kg), soil moisture (0.065×10^{18} kg), atmospheric liquid water content (0.0155×10^{18} kg), biosphere (0.002×10^{18} kg) and rivers (0.0017×10^{18} kg); [b]This number seems rather unrealistic, see text; 1 Allard (1995), 2 Brügmann and Matschullat (1997), 3 Chatterjee *et al.* (1995), 4 Chester (2000), 5 Damm *et al.* (1985a, b), 6 Das *et al.* (1995, 1996), 7 Donat and Bruland (1995), 8 Duce *et al.* (1991), 9 Gaillardet *et al.* (2003), 10 Heinrichs *et al.* (1997), 11 Kitano (1992), 12 Martin and Whitfield (1983), 13 Plant *et al.* (2005), 14 Pleßow *et al.* (1997a), 15 Reimann and Birke (2010), 16 Schleyer and Kerndorf (1992), 17 Whitfield (1979), 18 Ziegler and Gabriel (1997).

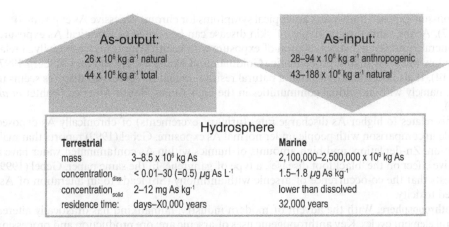

Figure 1.5. The hydrospheric As reservoir, related fluxes and residence time (see text).

concentrations within the plant above the standard upper limit of 1 mg kg^{-1} wet weight. Arsenic hyperaccumulators exist, e.g., certain types of grass, alfalfa and ferns that may show concentrations of 6–12 mg kg^{-1} wet weight at soil As concentrations of 25–50 mg kg^{-1} (Bhumbla and Keefer 1994). Extreme values up to several thousand mg As kg^{-1} were shown, e.g., by Ma *et al.* (2001, ▶ 13).

It is even more difficult to obtain reliable and representative data for animals. Therefore, mainly single values exist for a limited number of biota (Table 1.5; Fig. 1.1). Bowen (1979) compiled data for the human being; indicating a fictitious 'reference man' having a total of 18 mg As kg^{-1} body weight, equal to 0.26 mg As kg^{-1} with an average body weight of 70 kg. This appears high when using more recent numbers: Winter (2010) gives 0.05 mg As kg^{-1} as the standard human body concentration, equivalent to approximately 3.5 mg in a person of 70 kg body weight. Arsenic accumulates mainly in hair, skin, fingernails and toenails (Hutton 1987). In general, concentrations in these materials, as well as in blood and urine, are highly variable because of their relation to nutrient uptake. Compared to normal concentrations, elevated values occur in exposed populations, e.g., hair 0.04–32.5 mg As kg^{-1} and urine 5.8–170 µg As L^{-1} (Fergusson 1990; Table 1.6; ▶ 2, 14).

Arsenic exposure, via both air inhalation and oral intake of food and soil material, has a toxicological relevance for humans, animals and plants. Toxicity is orally acute and chronic as a neurotoxin and carcinogen (▶ 2). For humans, a tolerable resorbed dose is given with 0.3 mg As kg^{-1} day^{-1} as a Non-Observed Adverse Effect Level (NOAEL; Viereck-Götte and Ewers 1997).

The consumption of drinking water is the most relevant As intake pathway – approximately 30% of the As intake relates to drinking water (Appelo and Postma 2005). This is true both for the normal intake 0.04–1.4 mg day^{-1}, and for adverse toxic levels of 5–50 mg day^{-1} (lethal 50–340 mg day^{-1}; Bowen 1979). Other authors discuss levels of 1–5 mg day^{-1} for babies and 10–40 mg day^{-1} for adults as typical tolerable amounts (Pfannhauser and Widich 1979).

Legislation acknowledges the toxicological relevance with low drinking-water As thresholds with 10 µg L^{-1} (WHO standard; ▶ 2, 5). Inhalation may become an important intake-path for people with work-related exposure (Buat-Ménard *et al.* 1987), or can have local relevance. Goldschmidt (1958) points at the inhalation of toxic As amounts in swampy areas, where volatile arsine AsH$_3$ is being periodically liberated from anaerobic sites (Cheng and Focht 1979). Under non-chronical conditions, As retention within the human body can be counted in days. Human biomonitoring studies should, therefore, either be performed over longer periods of time or at different times using the same methods. Hyperkeratosis on the palms of the hands and soles of the feet, as well as disproportionate pigmentation of the skin on non-sun-exposed body parts are typical symptoms for chronic excessive As exposure (▶ 2, 14, 17). At the same time, all types of skin disease can be related to chronical As exposure. In general, however, a distinct relation of exposure with health symptoms – especially at relatively low rates – is highly problematic (Goldman and Dacre 1991; Guo and Valberg 1997; ▶ 2, 14). It also became known that a natural resistance against As-related diseases seems to exist, namely with individual communities in the high Andes, South America (Vahter *et al.* 1995).

This relates to higher As discharge rates (urinary, excrements) of chronically As-exposed people in comparison with people under normal As exposure. Gebel (1999) reports that malnutrition, Zn-depletion, and high amounts of humics within As contaminated water have a positive effect on the 'blackfoot disease', a type of gangrene. In the same paper, Gebel (1999) suggests that the co-occurrence of arsenic with antimony (Sb) leads to an alteration of As-related toxicity.

Anthroposphere. With the advent of modern industry, humankind has drastically altered natural element cycles. Key anthropogenic uses of arsenic are: ore production and processing (with melting and roasting in non-ferrous smelters, melting in iron works), high-temperature combustion (coal and oil burning power plants, waste incineration, cement works), wastes from intense husbandry (disinfectants, compost, dung, surplus-As from animal feed), household waste disposal, glassware production (decolouring agents), electronics industries (admixture in semiconductor production, arsenide as laser material to convert electrical energy into coherent light), metal treatment admixture (bronze production, lead and copper alloys), galvanising, ammunition factories (hardening and flight characteristics improvement of projectiles), chemistry (dyes and colours, wood preservatives, pesticides, pyrotechnics, drying agent for cotton, oil and dissolvent recycling), and pharmaceutical and medical products. These need to be considered when dealing with anthropogenic As fluxes (Han *et al.* 2003; Pacyna 1987; Reimann and de Caritat 1998; Savory and Wills 1984; Stoeppler 2004; Trueb 1996; Winter 2010). Han *et al.* (2003) claim that the cumulative global anthropogenic As production in the year 2000 was 4.53 million tons (As mining production > As generated from coal > As generated from petroleum). They conclude that within the "industrial age" the global input to the world arable soils has accumulated to 2.18 mg kg^{-1} – 1.2 times the contribution of the lithosphere. The data presented above do not fully support this hypothesis, but certainly agree that anthropogenic As fluxes make a major contribution in the global budget.

Between 20 and 35×10^6 kg (35×10^6 kg in 2002 and 20.26×10^6 kg in 2009) of elemental arsenic are being produced annually on a global level for the past 15 years (Brooks 2010). These numbers refer exclusively to the direct industrial use of the element and not to its anthropogenic turnover. Until the 1970s, approximately 80% of the produced arsenic was used in the manufacture of pesticides: simple inorganic As salts. For almost one century, As containing pesticides formed the backbone of the pesticide industry. Today, pesticides account for approximately 50% of the arsenic consumed, with organic As compounds now dominating the pesticide production (90%). Another 30% of the world As market relates to wood preservatives. The remaining uses include manufacture of glass, alloys, electronics, catalysts, fodder additions, and veterinary chemicals (Stoeppler 2004). In general, the As consumption in agriculture and lately in wood preservatives declines due to environmental regulations.

An estimated $28.4–94 \times 10^6$ kg As a^{-1} are introduced from the anthroposphere to pedosphere, hydrosphere and lithosphere through all sorts of waste materials (Chilvers and Peterson 1987; Fig. 1.6). Wastes may yield very high As concentrations (Heinrichs *et al.* 1997; Thein *et al.* 1997; Viereck-Götte and Herget 1997; Table 1.7). Analogous to the consequences of soil and water acidification from acidic deposition, the oxidation of sulphides leads to acidification and thus to increased As mobility, especially under reducing conditions (see hydrosphere in this chapter, and ▶ 11). Tailings from metal sulphide mining operations are well known and are substantial sources of both trace metal contamination (including arsenic) of ground and surface waters (Lottermoser 2010; ▶ 11). Mobilisation rates are directly related to those of

Table 1.6.   Natural median As concentrations in the biosphere (mg kg^{-1}, unless stated otherwise).

Medium	Concentration	Source
Plants, terrestrial[a]	<0.01–5 (0.1–0.65)	2, 6, 9
Mosses and lichens	0.22–0.26	11, 12
Bryophytes	0.2–7	1
Mushrooms, fungi	1.2–2.5	1
Equisetum	0.2	1
Ferns	1.3	1
Grasses	0.020–0.160	2
Wheat grain	0.010–0.070	2
Brown rice grains	0.110–0.200	2
Oat grains	0.010	2
Barley grains	0.003–0.018	2
Clover	0.280–0.330	2
Vegetables	0.01–1.5	1
Cabbage leaves	0.020–0.050	2
Kale	0.12	1
Carrots	0.040–0.080	2
Lettuce	0.020–0.250	2
Potatoes	0.030–0.200	2
Woody gymnosperms	0.02–1.2	1, 12
Woody angiosperms	0.02–2	1, 12
Spruce bark	10	11
Plants, aquatic	1–36	1
Algae	0.1–382	8, 9
Green algae	1.2–6	1
Red algae	6	1
Brown algae	8–30	1
Bacteria	0.1	1
Phytoplankton	12–36	1
Animals, terrestrial	very few data	
Arion ater (slugs)	0.014–1.17	10
Animals, marine	0.0036–166	3, 9
Annelida	6	1
Coelenterata	7.5–20	1, 3 wet wt
Crustacea	<0.1–270	1, 3, 8
Echinodermata	5–12.4	1, 3 wet wt
Mollusca	0.005–214	1, 3, 8
Pisces	0.05–450	1, 3, 8
Porifera	3.2–6.8	1, 3
Human beings	0.05–0.26	2, 16
Kidney	0.005–1.5	1, 2
Liver	0.02–1.6	1, 2
Lung tissue	0.078–0.141	4
Muscle tissue	0.009–0.65	1
Bones	0.08–1.6	1, 2
Hair	0.02–3.7	1, 2
Fingernails / toe nails	0.2–3	1
Milk	0.15–2.8 µg L^{-1}	4, 13
Blood	<0.5–10 mg L^{-1}	1, 2, 17
Urine	<1–8 µg L^{-1}	5, 7, 14, 15
Daily intake	63 µg person day^{-1}	4

[a]global As mass in plants: 0.184 × 10^6 kg (Markert 1992); 1 Bowen (1979), 2 Fergusson (1990: 386), 3 Francesconi *et al.* (1994), 4 Kabata-Pendias and Mukherjee (2007), 5 Krause *et al.* (1996), 6 Markert (1992), 7 Matschullat *et al.* (1999), 8 Neff (1997), 9 Onishi (1969), 10 Pozebon *et al.* (2008), 11 Reimann and de Caritat (1998), 12 Reimann *et al.* (2001), 13 Sternowski *et al.* (2002), 14 Vahter and Lind (1986), 15 White and Sabbioni (1998), 16 Winter (2010), 17 Yamauchi *et al.* (1992).

Table 1.7.   Arsenic fluxes (kg a^{-1}) and concentrations (mg kg^{-1}) in the anthroposphere.

Medium	Flux	Source
Waste release	28.4–94 × 10^6 kg a^{-1}	2
Coal and pretroleum incineration	23.9 × 10^6 kg a^{-1}	3
Annual extraction	20.26 × 10^6 kg (2009)	1
Medium	Concentration	Source
Wastes without metals	12 mg kg^{-1}	4
Wastes with 7% metals	13–18 mg kg^{-1}	4
Waste incineration ashes	1–45 mg kg^{-1}	7
Filter dusts from waste incineration	73–930 mg kg^{-1}	4, 6
Slag from waste incineration	30 mg kg^{-1}	3
Communal sewage sludge	4.7–6.5 mg kg^{-1}	4, 5
Building demolition material	1–59 mg kg^{-1}	7
Industrial side products, coal production tailings	9–21 mg kg^{-1}	7
Industrial side products, burnt	5–74 mg kg^{-1}	7
Coal incineration, granulate	1–7 mg kg^{-1}	7
Ash residue	1–46 mg kg^{-1}	7
Slags	1–8 mg kg^{-1}	7
Steel works slags	1–41 mg kg^{-1}	7
Non-ferrous smelter slags	5–1,035 mg kg^{-1}	7
Zinc smelters	33–470 mg kg^{-1}	7
Other smelter slags	8–85 mg kg^{-1}	7
Flue-gas desulfurisation salts	90 mg kg^{-1}	6
Electrofilter ash	560 mg kg^{-1}	6
As conc. increase with decreasing diameter of fly-ash	18.5 μm: 13.7 mg kg^{-1}; 2.4 μm: 132 mg kg^{-1}	3
NPK-fertilizer (Sweden)	0.59	5
P20-fertilizer (Sweden)	9.1	5
Solid pig manure (Sweden)	1.3 mg kg^{-1}	5

1 Brooks (2010), 2 Chilvers and Peterson (1987), 3 Han et al. (2003), 4 Heinrichs et al. (1997), 5 Kabata-Pendias and Muckerjee (2007), 6 Piver (1983), 6 Thein et al. (1997), 7 Viereck-Götte and Herget (1997).

sulphide mineral weathering (Appelo and Postma 2005). In Germany, this problem is less strongly related to base metal mining than to tailings and abandoned open pits from lignite mining (e.g., Kölling and Schulz 1993) and to coal mining waste heaps in the Ruhr area (Wiggering and Kerth 1991).

Hence, immobilisation techniques will continuously gain importance (e.g., Förstner 1994). Various methods exist today and show their reliability (▶ 3, 4). Schuiling (1996) discusses methods of the simultaneous removal of iron and arsenic from groundwater. Reduced groundwater will be oxidised when sprayed over sand filters (▶ 16). The two-valent iron then precipitates as amorphous Fe-hydroxide. The accompanying arsenite is co-precipitated as arsenate and forms Fe$^{(III)}$As$^{(V)}$O$_4$, a very stable complex under oxidising conditions. Examples from the Netherlands show the application of this technique. There, the As concentrations in a group of young students living on As-rich soils were lower in comparison to the control group, boys of the same age living on non-contaminated soils. At the same time, there are attempts to remediate As-contaminated sites that threaten ground and surface waters (▶ 16). Blumenroth and Bosecker (1999) proposed biological remediation of mine waste and tailings waters with CN-containing solutions (Lorösch 2001).

In summary, and while many data are available by now, there are still major shortcomings and serious knowledge gaps that need to be filled if global arsenic budgets are to be calculated with higher accuracy and precision. This is partly due to a lack of robust data, particularly for natural processes, e.g., volcanism and soil degassing, but certainly also relates to an obvious bias in studies that focus on As-contaminated areas on a very restricted scale. Starting with the lithosphere, it would be helpful to know its exact mass (within the usual limits). A factor 2 in the common estimates appears somewhat unsatisfactory. Nevertheless, the lithosphere is less of a problem when considering heuristic processes – except for volcanism. Mantle and crustal degassing data for arsenic and other components still are far from reliable. The shortest-term reservoir, the atmosphere, presents similar problems. Not nearly enough world-wide are available data to truly make a reliable assessment of atmospheric As masses on either hemisphere. Again, volcanic degassing contributes and related averaged data would be helpful, apart from the need for more representative data for As species. In principle, similar inconsistencies and knowledge gaps exist for both pedosphere and hydrosphere (the latter independent of freshwater or seawater reservoirs). While the data are considerably better than those from the atmosphere, there is much to be desired when compiling and calculating reservoir masses and global fluxes. The biosphere reservoir appears largely known. Yet, surprises such as the recent detection of hyperaccumulating plants and our rough estimates of global biospheric mass leave much to be desired. Last but certainly not least, we seem to face a rather large data gap in respect to As amounts in the anthroposphere. This stands in stark contrast with our human preoccupation with contamination and environmental problems. Our own backyard largely remains in the dark – we do not even have reliable estimates on how much arsenic there is (independent of species).

REFERENCES

Abernathy, C.O., Calderon, R.L. & Chappell, W.R.: *Arsenic exposure and health effects*. Chapman and Hall, London, 1997, p.429.

Allard, B.: Groundwater. In: Salbu, B. & Steinnes, E. (eds): *Trace elements in natural waters*. CRC Press, Boca Raton, 1995, pp.151–176.

Andreae, M.O.: Arsenic in rain and the atmospheric mass balance of arsenic. *J. Geophys. Res.* 85:C8 (1980), pp.4512–4518.

Andreas, H.: *Schweinfurter Grün – das brillante Gift*, 1996. Available at http://de.wikipedia.org/wiki/Chemie_in_unserer_Zeit 30: pp.23–31.

Appelo, C.A.J. & Postma, D.: *Geochemistry, groundwater and pollution*. 2nd ed. Balkema, Rotterdam, 2005, p.678.

Azcue, J.M. & Nriagu, J.O.: Arsenic: historical perspectives. In: Nriagu, J.O. (ed): *Arsenic in the environment*. Part 1: cycling and characterization. *Adv. Environ. Sci.*, Volume 26. John Wiley & Sons, New York, 1994, pp.1–15.

Bentley, R. & Chasteen, T.H.: Microbial methylation of metalloids: arsenic, antimony and bismuth. *Microbiol. Mol. Biol. R.* 66:2 (2002), pp.250–271.

Berner, E.K. & Berner, R.A.: *Global environment – water, air, and geochemical cycles*. Prentice Hall, Upper Saddle River, 1996, p.376.

Bhattacharya, P., Mukherjee, A.B., Bundschuh, J., Zevenhoven, R. & Loeppert, R.H.: Arsenic in soil and groundwater environment. Biogeochemical interactions, health effects and remediation. In: Nriagu, J.O. (ser ed): *Trace metals and other contaminants in the environment*, Volume 9. Elsevier, Amsterdam, 2007, p.653.

Bhumbla, D.K. & Keefer, R.F.: Arsenic mobilization and bioavailability in soils. In: Nriagu, J.O. (ed): *Arsenic in the environment*. Part 1: cycling and characterization. *Adv. Environ. Sci.*, Volume 26. John Wiley & Sons, New York, 1994, pp.51–82.

Blumenroth, P. & Bosecker, K.: Mikrobieller Abbau von Cyanid in Prozeßwässern der Goldgewinnung. In: Wippermann, T. (ed): *Bergbau und Umwelt – langfristige geochemische Einflüsse*. GUG Schriftenreihe Geowissenschaften + Umwelt, Volume 5. Springer, Berlin, 1999, pp.183–197.

Borba, R., Figueiredo, B.R. & Matschullat, J.: Geochemical distribution of arsenic in waters, sediments and weathered gold-mineralized rock from Iron Quadrangle, Brazil. *Environ. Geol.* 44:1 (2003), pp.39–52.

Bowen, H.J.M.: *Environmental chemistry of the elements.* Academic Press, London, 1979, p.348.

Brooks, W.E.: Arsenic. World production and reserves. USGS Mineral Resource database, 2010. Available at http://minerals.usgs.gov/minerals/pubs/commodity/arsenic/mcs-2010-arsen.pdf.

Brügmann, L. & Matschullat, J.: Zur Biogeochemie und Bilanzierung von Schwermetallen in der Ostsee. In: Matschullat, J., Tobschall, H.J. & Voigt, H.J. (eds): *Geochemie und Umwelt. Relevante Prozesse in Atmo-, Pedo- und Hydrosphäre.* Springer, Berlin, 1997, pp.267–290.

Bruland, K.W.: Trace elements in sea water. In: Riley, J.P. & Chester, R. (eds): *Chemical oceanography,* Volume 8. Academic Press, London, 1983, pp.157–220.

Buat-Ménard, P., Peterson, P.J., Havas, M., Steinnes, E. & Turner, D.: Group report: Arsenic. In: Hutchinson, T.C. & Meema, K.M. (eds): *Lead, mercury, cadmium and arsenic in the environment.* SCOPE, Volume 31. John Wiley & Sons, New York, 1987, pp.43–50.

Bundschuh, J., Bhattacharya, P. & Chandrasekharam, D. (eds): *Natural arsenic in groundwater. Occurrence, remediation and management.* Balkema, Leiden, 2005, p.339.

Bundschuh, J., Armienta, M.A., Bhattacharya, P., Matschullat, J., Birkle, P. & Rodriguez, R. (eds): *Natural arsenic in groundwaters of Latin America* – Abstract volume. Mexico-City, 2006, p.89.

Bundschuh, J., Pérez Carrera, A. & Litter, M. (eds): *Distribución del arsénico en las regiones Ibérica e Iberoamericana.* CYTED Argentina, 2008, p.230.

Bundschuh, J., Armienta, M.A., Birkle, P., Bhattacharya, P., Matschullat, J. & Mukherjee, A.B. (eds): *Natural arsenic in groundwaters of Latin America,* Volume 1. Balkema, Amsterdam, 2009, p.742.

Calmano, W.: Probleme mit Hamburger Hafenschlick. In: Lozan, J.L. & Kausch, H. (eds): *Warnsignale aus Flüssen und Ästuaren.* Berlin, Paul Parey, 1996, pp.124–129.

Chappell, W.R., Abernathy, C.O. & Calderon, R.L. (eds): *Arsenic exposure and health effects,* Volume III. Elsevier, Amsterdam, 1999, p.416.

Chappell, W.R., Abernathy, C.O. & Calderon, R.L. (eds): *Arsenic exposure and health effects,* Volume IV. Elsevier, Amsterdam, 2001, p.467.

Chappell, W.R., Abernathy, C.O., Calderon, R.L. & Thomas, D.J. (eds): *Arsenic exposure and health effects,* Volume V. Elsevier, Amsterdam, 2003, p.533.

Chatterjee, A., Das, D., Mandal, B.K., Chowdhury, T.R., Samanta, G. & Chakraborti, D.: Arsenic in groundwater in six districts of West Bengal, India: the biggest arsenic calamity in the world. I: arsenic species in drinking water and urine of affected people. *Analyst.* 120:3 (1995), pp.643–650.

Cheng, C.N. & Focht, D.D.: Production of arsine and methylarsine in soil and in culture. *Appl. Environ. Microbiol.* 38 (1979), pp.494–498.

Chester, R.: *Marine geochemistry.* 2nd ed.: Chapman & Hall, London, 2000, p.698.

Chilvers, D.C. & Peterson, P.J.: Global cycling of arsenic. In: Hutchinson, T.C. & Meema, K.M. (eds): *Lead, mercury, cadmium and arsenic in the environment.* SCOPE, Volume 31. John Wiley & Sons, New York, 1987, pp.279–301.

Cullen, W.R.: *Is arsenic an aphrodisiac? The sociochemistry of an element.* RSC Publishing, Cambridge, UK, 2008, p.412.

Cullen, W.R. & Reimer, K.J.: Arsenic speciation in the environment. *Chem. Rev.* 89:4 (1989), pp.713–764.

Damm, K.L. von Edmond, J.M., Grant, B., Measures, C.I., Walden, B. & Weiss, R.F.: Chemistry of submarine hydrothermal solutions at 218N, East Pacific Rise. *Geochim. Cosmochim. Acta* 49:11 (1985a), pp.2197–2220.

Damm, K.L. von Edmond, J.M., Measures, C.I. & Grant, B.: Chemistry of submarine hydrothermal solutions at Guaymas basin, Gulf of California. *Geochim. Cosmochim. Acta* 49:11 (1985b), pp.2221–2237.

Däßler, H.G.: *Einfluß von Luftverunreinigungen auf die Vegetation. Ursachen, Wirkungen, Gegenmaßnahmen.* 3rd ed: Gustav Fischer, Jena, 1986, p.223.

Das, D., Chatterjee, A., Mandal, B.K., Samanta, G. & Chakraborti, D.: Arsenic in groundwater in six districts of West Bengal, India: the biggest arsenic calamity in the world. 2. Arsenic concentration in drinking water, hair, nails, urine, skin-scale and liver tissue biopsy of the affected people. *Analyst* 120:3 (1995), pp.917–924.

Das, D., Samanta, G., Mandal, B.K., Chowdhury, T.R., Chanda, C.R., Chowdhury, P.P., Basu, G.K. & Chakraborti, D.: Arsenic in groundwater in six districts of West Bengal, India. *Environ. Geochem. Health* 18:1 (1996), pp.5–15.

Donat, J.R. & Bruland, K.W.: Trace elements in the ocean. In: Salbu, B. & Steinnes, E. (eds): *Trace elements in natural waters*. Boca Raton, CRC Press, 1995, pp.247–281.

Driehaus, W.: Arsenentfernung mit Mangandioxid und Eisenhydroxid in der Trinkwasseraufbereitung. *VDI-Reports* 133:15 (1994), p.117.

Duce, R.A., Liss, P.S., Merrill, J.T., Atlas, E.L., Buat-Ménard, P., Hicks, B.B., Miller, J.M., Prospero, J.M., Arimoto, R., Church, T.M., Ellis, W., Galloway, J.N., Hansen, L., Jickells, T.D., Knap, A.H., Reinhard, K.H., Schneider, B., Soudine, A., Tokos, J.J., Tsunogai, S., Wollast, R. & Zhou, M.: The atmospheric input of trace species to the world ocean. *Global Biogeochem. Cycles* 5:3 (1991), pp.193–259.

Ebdon, L., Pitts, L., Cornelis, R., Crews. H., Donard, O.F.X. & Quevauviller, P. (eds): *Trace element speciation for environment, food and health*. Royal Soc. Chem., Cambridge, 2001, p.391.

Feldmann, J., Devalla, S., Raab, A. & Hansen, H.R.: Analytical strategies for arsenic speciation in environmental and biological samples. In: Hirner, A.V. & Emons, H. (eds): Organic metal and metalloid species in the environment. *Analysis, distribution, preocesses and toxicological evaluation*. Springer, Heidelberg, 2004, pp.41–70.

Ferguson, J.F. & Gavis, J.: A review of the arsenic cycle in natural waters. *Water Res.* 6:11 (1972), pp.1259–1274.

Fergusson, J.E.: *The heavy elements: chemistry, environmental impact and health effects*. Pergamon Press, Oxford, 1990, p.614.

Fiedler, H.J. & Rösler, H.J. (eds): *Spurenelemente in der Umwelt*. Jena, Gustav Fischer, 1987, p.278.

Finkelman, R.B., Belkin, H.E. & Zheng, B.: Health impacts of domestic coal use in China. *PNAS* 96:7 (1999), pp.3427–3431.

Förstner, U.: Geochemische Konzepte in Abfallforschung und -praxis. In: Matschullat, J. & Müller, G.: *Geowissenschaften und Umwelt*. Berlin, Springer, 1994, pp.315–326.

Fowler, B.A. (ed): Biological and environmental effects of arsenic. *Topics in environmental health*, Volume 6. Elsevier, Amsterdam, 1983, p.281.

Francesconi, K.A., Edmonds, J.S. & Morita, M.: Biotransformation of arsenic in the marine environment. In: Nriagu, J.O. (ed): *Arsenic in the environment*. 1. Cycling and characterization. *Adv. Environ. Sci. Technol.*, Volume 26. John Wiley & Sons, New York, 1994, pp.221–261.

Gaillardet, J., Viers, J. & Dupré, B.: Trace elements in river waters. In: Drever, J.I. (ed): *Surface and groundwater, weathering and soils*. In: Holland, H.D. & Turekian, K.K. (ser eds): *Treatise on geochemistry*, Volume 5. Elsevier, Amsterdam, 2003, pp.225–272.

Gebel, T.W.: Arsenic and drinking water contamination. *Science* 283:5407 (1999), pp.1458–1459.

God, R.: Geogene Arsengehalte außergewöhnlichen Ausmaßes in Böden, nördliche Saualpe – ein Beitrag zur Diskussion um Grenzwerte von Spurenelemente in Boden. *Berg- und Hüttenmännische Monatshefte*, Volume 139:12. Leoben, 1994, pp.442–449.

God, R. & Heiss, G.: Die Arsenanomalie Feistritz am Wechsel Niederösterreich. *Jahrb. Geol. Bundesanstalt Wien* 139:4 (1996), pp.437–444.

Goldman, M. & Dacre, J.C.: Inorganic arsenic compounds: are they carcinogenic, mutagenic, teratogenic? *Environ. Geochem. Health* 13:4 (1991), pp.179–191.

Goldschmidt, V.M.: The principles of distribution of chemical elements in minerals and rocks. *J. Chem. Soc. London* 74 (1937), pp.655–673.

Goldschmidt, V.M.: Geochemistry. In: Muir, A. (ed): Oxford University Press, Oxford, 1958, pp.468–475.

Guo, H.R., Valberg, P.A.: Evaluation of the validity of the US EPA' cancer risk assessment of arsenic for low level exposures: a likelihood ratio approach. *Environ. Geochem. Health* 19:4 (1997), pp.133–141.

Han, F.X., Su, Y., Monts, D.L., Plodinec, M.J., Banin, A. & Triplett, G.E.: Assessment of global industrial-age anthropogenic arsenic contamination. *Naturwiss.* 90:9 (2003) pp.395–401.

Harrison, R.M. & Mora, S.J. de.: Introductory chemistry for the environmental sciences. 2nd ed; *Cambridge Environ. Chem. Ser.* Volume 7. Cambridge University Press, Cambridge, 1996, p.373.

Heinrichs, H. & Brumsack, H.J.: Anreicherung von umweltrelevanten Metallen in atmosphärisch transportierten Schwebstäuben aus Ballungszentren. In: Matschullat, J., Tobschall, H.J., & Voigt, H.J. (eds): *Geochemie und Umwelt. Relevante Prozesse in Atmo-, Pedo- und Hydrosphäre*. Berlin, Springer, 1997, pp.25–36.

Heinrichs, H., Hundesrügge, T. & Brumsack, H.J.: Siedlungsabfälle: Verwertung, Verbrennung, Deponierung. In: Matschullat, J., Tobschall, H.J. & Voigt, H.J. (eds): *Geochemie und Umwelt. Relevante Prozesse in Atmo-, Pedo- und Hydrosphäre*. Berlin, Springer, 1997, pp.189–202.

Hong, J., Calmano, W. & Förstner, U.: Interstitial waters. In: Salbu, B. & Steinnes, E. (eds): *Trace elements in natural waters*. CRC Press, Boca Raton, 1995, pp.117–150.

Hutchinson, T.C. & Meema, K.M. (eds): *Lead, mercury, cadmium and arsenic in the environment.* SCOPE, Volume 31. John Wiley & Sons, New York, 1987, p.384.

Hutton, M.: Human health concerns of lead, mercury, cadmium and arsenic. In: Hutchinson, T.C. & Meema, K.M. (eds): Lead, mercury, cadmium and arsenic in the environment. *SCOPE*, Volume 31. John Wiley & Sons, New York, 1987, pp.53–6.

Imai, N., Terashima, S., Ohta, A., Mikoshiba, M., Okai, T., Tachibana, Y., Togashi, S., Matsuhita, Y., Kanai, Y., Kamioka, H. & Taniguchi, M.: Geochemical map of Japan. Geol. Survey Japan, AIST, Tsukuba, 2004, p.209.

Jacob, D.J.: *Introduction to atmospheric chemistry.* Princeton Univ. Press, Princeton, 1999, p.264.

Jekel, M.R.: Removal of arsenic in drinking water treatment. In: Nriagu, J.O. (ed): *Arsenic in the environment.* 1: cycling and characterisation. *Adv. Environ. Sci.* Volume 26. John Wiley & Sons, New York, 1994, pp.119–132.

Kabata-Pendias, A. & Mukherjee, A.B.: *Trace elements from soil to human.* Springer, Berlin, 2007, p.550.

Kitano, Y.: Water chemistry. In: Nierenberg, W.A. (ed): *Encyclopedia of Earth system science*, Volume 4. Academic Press, San Diego, 1992, pp.449–470.

Koljonen, T. (ed): *Geochemical atlas of Finland.* Geol Survey Finland, Espoo, p.2 Till 218.

Kölling, M. & Schulz, H.D.: Pyritverwitterung und saure Grubenwässer in Halden des Braunkohlentagebaus. In: Dörhöfer, G., Thein, J. & Wiggering, H. (eds): Abfallbeseitigung und Deponien – Anforderungen an Abfall und Deponie. *Umweltgeologie heute,* Volume 1. Ernst & Sohn, Berlin, 1993, pp.41–47.

Krause, C., Babisch, W., Becker, K., Bernigau, W., Helm, D., Hoffmann, K., Nollke, P., Schulz, C., Schwalbe, R., Seifert, T. & Thefeld, W. (1996) Umweltsurvey 1990/92. Available at http://www.umweltbundesamt.de/gesundheit/survey/us9092/urin.htm.

Lantzy, R.J. & Mackenzie, F.T.: Atmospheric trace metals: global cycles and assessment of man's impact. *Geochim. Cosmochim. Acta* 43:4 (1979), pp.511–525.

Lide, D.R. (ed): *CRC Handbook of chemistry and physics.* 77th ed. CRC Press, Boca Raton, 1996.

Lin Zhao, L., Bao, S. & Cheng, Z.: The uses of arsenic in traditional Chinese medicine. In SEGH, (ed): *Third international conference on arsenic exposure and health effects.* Book of abstracts, San Diego, 12–15 July, 1998, p.148.

Lindqvist, O., Johansson, K., Aastrup, M., Andersson, A., Bringmark, L., Hovsenius, G., Håkanson, L., Iverfeldt, Å., Meili, M. & Timm, B.: Mercury in the Swedish environment – recent research on causes, consequences and corrective methods. *Water Air Soil Pollut.* 55:1 (1991), p.261.

Lorösch, J.: Process and environmental chemistry of cyanidation. Degussa AG, Frankfurt, 2001, p.504.

Lottermoser, B.G.: *Mine wastes. Characterization, treatment and environmental impacts.* 3rd ed., Springer, Heidelberg, 2010, p.400.

Ma, L.Q., Komar, K.M., Tu, C., Zhang, W., Cai, Y. & Kennelly, E.D.: A fern that hyperaccumulates arsenic. *Nature* 409:6820 (2001), p.579.

Mackenzie, F.T., Lantzy, R.J. & Paterson, V.: Global trace metal cycles and predictions. *J. Internat. Assoc. Math. Geol.* 11:2 (1979), pp.99–142.

Mandal, B.K. & Suzuki, K.T.: Arsenic around the world: a review. *Talanta* 58:1 (2002), pp.201–235.

Markert, B.: Presence and significance of naturally occurring chemical elements of the periodic system in the plant organism and consequences for future investigations on inorganic environmental chemistry ecosystems. *Vegetatio./Plant Ecol.* 103:1 (1992), pp.1–30.

Martin, J.M. & Whitfield, M.: The significance of the river input of chemical elements to the ocean. In: Wong, C.S., Boyle, E., Bruland, K.W., Burton, J.D. & Goldberg, E.D. (eds): *Trace metals in sea water.* Plenum Press, New York, 1983, pp.265–296.

Matschullat, J.: Heavy metal contamination of soils: reuse versus disposal. In: Reuther, R. (ed): *Geochemical approaches to environmental engineering of metals.* Springer, Berlin, 1996, pp.81–88.

Matschullat, J.: Arsenic in the geosphere – a review. *Sci. Total Environ.* 249:1–3 (2000), pp.297–312.

Matschullat, J. & Kritzer, P.: Atmosphärische Deposition von Spurenelementen in 'Reinluftgebieten'. In: Matschullat, J., Tobschall, H.J. & Voigt, H.J. (eds): *Geochemie und Umwelt. Relevante Prozesse in Atmo-, Pedo- und Hydrosphäre.* Springer, Berlin, 1997, pp.3–24.

Matschullat, J., Bozau, E., Brumsack, H.J., Fänger, R., Halves, J., Heinrichs, H., Hild, A., Lauterbach, G., Leßmann, D., Schaefer, M., Schneider, J., Schubert, M. & Sudbrack, R.: Stoffdispersion Osterzgebirge – Ökosystemforschung in einer alten Kulturlandschaft. In: Matschullat, J. & Müller, G. (eds): *Geowissenschaften und Umwelt.* Springer, Berlin, 1994, pp.227–242.

Matschullat, J., Müller, G., Naumann, U. & Schilling, H.: Hydro- und Sedimentgeochemie im Ein-
zugsgebiet der schwarzen Elster – Aus dem Verbundprojekt 'Elbe-Nebenflüsse' des BMBF, Phase II.
Heidelberger Beitr Umwelt-Geochem. 10 (1997), p.100.

Matschullat, J., Scharnweber, T., Garbe-Schönberg, D., Walther, A. & Wirth, V.: Epilithic lichens –
atmospheric deposition monitors of trace elements and organohalogens? *J. Air Waste Manage.
Assoc.* 49:10 (1999), pp.174–184.

Matschullat, J., Borba, R.P., Deschamps, E., Figueiredo, B.R., Gabrio, T. & Schwenk, M.: Human
and environmental contamination in the Iron Quadrangle, Brazil. *Appl. Geochem.* 15:2 (2000),
pp.181–190.

McCleskey, R.B., Nordstrom, D.K. & Maest, A.S.: Preservation of water samples for arsenic (III/V)
determinations: an evaluation of the literature and new analytical results. *Appl. Geochem.* 19 (2004),
pp.995–1009.

Milliman, J.D.: Transfer of river-borne particulate material to the oceans. In: Martin, J.M., Burton, J.D.
& Eisma, D. (eds): *River inputs to the oceans.* March 1979. Proc review workshop at FAO headquar-
ters Rome, Italy, 1980, pp.5–12.

Milliman, J.D. & Meade, R.H.: World-wide delivery of river sediment to the oceans. *J. Geol.* 91:1 (1983),
pp.1–21.

Müller, G.: Schwermetalle und organische Schadstoffe in den Flußsedimenten. In: Lozan, J.L. & Kausch,
H. (eds): *Warnsignale aus Flüssen und Ästuaren.* Paul Parey, Berlin, 1996, pp.113–123.

Mukherjee, A.B., Bhattacharya, P., Sajwan, K., Zevenhoven, R. & Matschullat, J.: Global arsenic and
antiomony flow through coal and their cycling in groundwater environment. In: Bhattacharya, P.,
Ramanathan, A.L., Mukherjee, A.B., Bundschuh, J., Chandrasekharam, D. & Keshari. A.K. (eds):
Groundwater for sustainable development: problems, perspectives and challenges. Taylor & Francis,
London, 2008, pp.323–334.

Neff, J.M.: Ecotoxicology of arsenic in the marine environment. *Environ. Toxicol. Chem.* 16:5 (1997),
pp.917–927.

Neff, J.M.: Arsenic in the ocean. In: Jeff, J.M. (ed); *Bioaccumulation in marine organisms. Effect of contaminants from oil-well produced water.* Elsevier, Amsterdam, 2002, pp.57–78.

Ng, J.C., Wang, J. & Shraim, A.: A global health problem caused by arsenic from natural sources. *Chem-
osphere* 52:9 (2003), pp.1353–1359.

Nriagu, J.O. & Pacyna, J.M.: Quantitative assessment of worldwide contamination of air, water and
soils by trace metals. *Nature* 333:6196 (1998), pp.134–139.

Nriagu, J.O.: Heavy metal pollution poisoning the biosphere? *Environment* 32:7 (1990), pp. 6–11.

Nriagu, J.O. (ed) *Arsenic in the environment.* 1. Cycling and characterization. *Adv. Environ. Sci.* Volume 26.
John Wiley & Sons, New York, 1994, p.430.

Onishi, H.: Arsenic. In: Wedepohl, K.H. (ed): *Handbook of geochemistry.* Volume II, 3. Springer, Berlin,
1969, p.33.

Pacyna, J.M.: Atmospheric emissions of arsenic, cadmium, lead and mercury from high temperature
processes in power generation and industry. In: Hutchinson, T.C. & Meema, K.M. (eds): Lead, mercury, cadmium and arsenic in the environment. *SCOPE*, Volume 31 John Wiley & Sons, New York,
1987, pp.69–87.

Pfannhauser and Widich (1979), cited in Fergusson, (1985).

Piver, W.T.: Mobilization of arsenic by natural and industrial processes. In: Fowler, B.A. (ed): *Biological
and environmental effects of arsenic.* Elsevier, Amsterdam, 1983, pp.1–50.

Planer-Friedrich, B., Lehr, C., Matschullat, J., Merkel, B.J., Nordstrom, D.K. & Sandstrom, M.W.: Speciation of volatile arsenic at geothermal features in Yellowstone National Park. *Geochim. Cosmochim.
Acta* 70:10 (2006), pp.2480–2491.

Planer-Friedrich, B., London, J., McCleskey, R.N., Nordstrom, D.K. & Wallschlager, D. Thioarsenates
in geothermal waters of Yellowstone National Park: determination, preservation, and geochemical
importance. *Environ. Sci. Technol.* 41:15 (2007), pp.5245–5251.

Plant, J., Kinniburgh, D.G., Smedley, P.L., Fordyce, F.M. & Klinck, B.A.: Arsenic and selenium. In:
Sherwood Lollar, B. (ed): *Environmental geochemistry.* In: Holland, H.D., Turekian, K.K. (eds):
Treatise on geochemistry, Volume 9 Elsevier, Amsterdam, 2005, pp.17–66.

Pleßow, A., Bielert, U., Heinrichs, H. & Steiner, I.: Problematik der Grundwasserversauerung und das
Lösungsverhalten von Spurenstoffen. In: Matschullat, J., Tobschall, H.J., Voigt, H.J. (eds): *Geochemie und Umwelt. Relevante Prozesse in Atmo-, Pedo- und Hydrosphäre.* Springer, Berlin, (1997a),
pp.395–408.

Pleßow, A., Pleßow, K. & Heinrichs, H.: Schadstoffbelastung von Straßenkehrreicht und Sedimenten der Regenwasserkanalisation durch den Straßenverkehr am Beispiel von Göttingen. Z. Umweltchem. Ökotox. 9:6 (1997b), pp.353–354.

Plessow, K., Acker, K., Heinrichs, H., & Möller, D.: Time study of trace elements and major ions during two cloud events at the Mt. Brocken. Atmos. Environ. 35:2 (2001), pp.367–378.

Pozebon, D., Dressler, V.L., Becker, J.S., Matusch, A., Zoriy, M. & Becker, J.S.: Biomonitoring of essential and toxic elements in small biological tissues by ICP-MS. J. Anal. Atom. Spectrom. 23:9 (2008), pp.1281–1284.

Preziosi, E., Giuliano, G. & Vivona, R.: Natural background levels and threshold values derivation for naturally As, V and F-rich water bodies: a methodological case study in Central Italy. Environ. Earth Sci. 61:5 (2010), pp.885–897.

Rahn, K.A.: The chemical composition of the atmospheric aerosol. Technical Report, Graduate School of Oceanography, Univ Rhode Island, Kingston, 1976; cited in Pacyna (1987).

Reimann, C. & de Caritat, P.: Chemical elements in the environment. Springer, Berlin, 1998, p.398.

Reimann, C. & de Caritat, P.: Intrinsic flaws of element enrichment factors (EFs) in environmental geochemistry. Environ. Sci. Technol. 34:24 (2000), pp.5084–5091.

Reimann, C. & de Caritat, P.: Distinguishing between natural and anthropogenic sources for elements in the environment: Regional geochemical surveys versus enrichment factors. Sci. Total Environ. 337:1–3 (2005), pp.91–107.

Reimann, C. & Birke, M. (eds): Geochemistry of European bottled water. Borntraeger Science, Stuttgart, 2010, p.268.

Reimann, C., Koller, F., Frengstad, B., Kashulina, G., Niskavaara, H. & Englmaier, P.: Comparison of the element composition in several plant species and their substrate from a 1,500,000-km² area in Northern Europe. Sci. Total Environ. 278:1–3 (2001), pp.87–112.

Reimann, C., Siewers, U., Tarvainen, T., Bityukova, L., Eriksson, J., Gilucis, A., Gregorauskiene, V., Lukashev, V.K., Matinian, N.N., Pasieczna, A.: Agricultural soils in northern Europe: a geochemical atlas. Geol. Jb. D., SD5: (2003), p.279.

Reimann, C., Matschullat, J., Birke, M. & Salminen, R.: Arsenic distribution in the environment: the effects of scale. Appl. Geochem. 24:7 (2009), pp.1147–1167.

Riederer, J.: Archäologie und Chemie – Einblicke in die Vergangenheit. Staatliche Museen Preußischer Kulturbesitz und Rathgen-Forschungslabor, Berlin, 1987, p.276.

Rühling, A.: Atmospheric heavy metal deposition in Europe – estimations based on moss analysis. Nord. 8 (1994), p.53.

Rühling, A. & Steinnes, E.: Atmospheric heavy metal deposition in Europe 1995–1996. Nord. 15 (1998), p.66.

Rudnick, R.L. & Gao, S.: Composition of the continental crust. In: Rudnick, R.L. (ed): The crust. In: Holland, H.D. & Turekian, K.K. (ser eds): Treatise on geochemistry, Volume 3, 2004, pp.1–64.

Sachs, R.M. & Michael, J.L.: Comparative phytotoxicity among four arsenical herbicides. Weed. Sci. 19:5 (1971), pp.558–564.

Sadiq, M.: Arsenic chemistry in soils: An overview of thermodynamic predictions and field observations. Water Air Soil Pollut. 93:1–4 (1997), pp.117–136.

Salminen, R., Chekushin, V., Tenhola, M., Bogatyrev, I., Glavatskikh, S.P., Fedotova, E., Gregorauskiene, V., Kashulina, G., Niskavaara, H., Polischuok, A., Rissanen, K., Seleno, L., Tomilina, O. & Zhdanova, L.: Geochemical atlas of eastern Barents region. Elsevier, Amsterdam, 2004, p.548.

Salminen, R., Batista, M.J., Bidovec, M., Demetriades, A., De Vivo, B., De Vos, W., Duris, M., Gilucis, A., Gregorauskiene, V., Halamic, J., Heitzmann, P., Lima, A., Jordan, G., Klaver, G., Klein, P., Lis, J., Locutura, J., Marsina, K., Mazreku, A., O'Connor, P.J., Olsson, S.Å., Ottesen, R.T., Petersell, V., Plant, J.A., Reeder, S., Salpeteur, I., Sandström, H., Siewers, U., Steenfelt, A. & Tarvainen, T. (eds): Geochemical atlas of Europe. 1. Background information, methodology and maps, Geol. Survey Finland, Espoo, 2005, p.526.

Savory, J. & Wills, M.R.: Arsen. In: Merian, E. (ed); Metalle in der Umwelt. Weinheim: VCH, 1984, pp.319–334.

Schleyer, R. & Kerndorff, H.: Die Grundwasserqualität westdeutscher Trinkwasserressourcen. Weinheim: VCH Verlagsgesellschaft, 1992, p.257.

Schröter, W., Lautenschlager, K.H. & Bibrack, H. (eds): Taschenbuch der Chemie. Harri Thun, Frankfurt, 1983, pp.83–84.

Schuiling, R.D.: Geochemical engineering: principles and case studies. In: Reuther, R. (ed): Geochemical approaches to environmental engineering of metals. Springer, Berlin, 1996, pp.3–12.

Schulte, A. & Blum, W.E.H.: Schwermetalle in Waldökosystemen. In: Matschullat, J., Tobschall, H.J. & Voigt, H.J. (eds): *Geochemie und Umwelt. Relevante Prozesse in Atmo-, Pedo- und Hydrosphäre*. Springer, Berlin, 1997, pp.53–74.

SEGH (ed): *Third international conference on arsenic exposure and health effects*. Book of abstracts. San Diego, Ca, 12–15 July, p.173.

Siewers, U. & Herpin, U.: Schwermetalleinträge in Deutschland: Moos-Monitoring 1995/96. *Geol. Jb. DSD*. 2 (1998), p.199.

Smedley, P.L. & Kinniburgh, D.G.: A review of the source, behaviour and distribution of arsenic in natural waters. *Appl. Geochem*. 17:5 (2002), 517–568.

Smith, A.H., Hopenhayn-Rich, C., Bates, M.N., Goeden, H.M., Hertz-Picciotto, I., Duggan, H.M., Wood, R., Kosnett, M.J. & Smith, M.T.: Cancer risks from arsenic in drinking water. *Environ. Health Perspectives* 97 (1992), pp.259–267.

Spini, G., Profumo, A., Riolo, C., Dalla Stella, C. & Zecca, E.: Speciation of arsenic in the atmosphere. *Toxicol. Environ. Chem*. 46:1–2 (1994), pp.81–95.

Sternowski, H.J., Moser, B. & Szadkowsky, D.: Arsenic in breast milk during the first 3 months of lactation. *Int. J. Hyg. Environ. Health* 205:5 (2002), pp.405–409.

Stoeppler, M.: Arsenic. In: Merian, E., Anke, M., Ihnat, M. & Stoeppler, M. (eds): *Elements and their compounds in the environment. Occurrence, analysis and biological relevance*. 2nd ed., Volume 3. Wiley-VCH, Weinheim, 2004, pp.1321–1364.

Tamaki, S. & Frankenberger, W.T. jr.: Environmental biochemistry of arsenic. *Rev. Environ. Contam. Toxicol*. 124:1 (1992), pp.79–110.

Taylor, S.R. & McLennan, S.M.: The geochemical evolution of the continental crust. *Rev. Geophys*. 33:2 (1995), pp.241–265.

Thein, J., Veerhoff, M. & Klinger, C.: Geochemische Barrieren bei Versatzbergwerken im Fels. In: Matschullat, J., Tobschall, H.J. & Voigt, H.J. (eds): *Geochemie und Umwelt. Relevante Prozesse in Atmo-, Pedo- und Hydrosphäre*. Springer, Berlin, 1997, pp.227–244.

Toksoz, M.N.: The subduction of the lithosphere. In: Wilson, J.T. (ed); *Continents adrift and continents aground. San Francisco: Scientific American*; WH Freeman and Company, Volume 9, 1976, pp.113–122.

Trueb, L.F.: *Die chemischen Elemente – ein Streifzug durch das Periodensystem*. Hirzel, Stuttgart, 1996, pp.300–305; p.434.

Vahter, M. & Lind, B.: Concentration of arsenic in urine of the general population in Sweden. *Sci. Total Environ*. 54:1 (1986), pp.1–12.

Vahter, M., Concha, G., Nermell, B., Nilsson, R., Dulout, F. & Natarajan, A.T.: A unique metabolism of inorganic arsenic in native Andean women. *Eur. J. Pharmacol*. 293:4 (1995), pp.455–462.

Viereck-Götte, L. & Ewers, U.: Grundlagen und Verfahren der Ableitung von Richtwerten. In: Matschullat, J., Tobschall, H.J. & Voigt, H.J. (eds): *Geochemie und Umwelt. Relevante Prozesse in Atmo-, Pedo- und Hydrosphäre*. Springer, Berlin, 1997, pp.245–264.

Viereck-Götte, L. & Herget, J.: Zur Geochemie der Boden industriell geprägter urbaner Gebiete. In: Matschullat, J., Tobschall, H.J. & Voigt, H.J. (eds): *Geochemie und Umwelt. Relevante Prozesse in Atmo-, Pedo- und Hydrosphäre*. Springer, Berlin, 1997, pp.127–150.

Walsh, P.R., Duce, R.A. & Fasching, J.L.: Tropospheric arsenic over marine and continental regions. *J. Geophys. Res*. 84:C4 (1979), pp.1710–1718.

Wedepohl, K.H.: The composition of the continental crust. *Geochim. Cosmochim. Acta* 59:7 (1995), pp.1217–1232.

Wedepohl, K.H.: The composition of Earth's upper crust, natural cycles of elements, natural resources. In: Merian, E., Anke, M., Ihnat, M. & Stoeppler, M. (eds): *Elements and their compounds in the environment. Occurrence, analysis and biological relevance*. 2nd ed., Volume 1, 2004, pp.3–16.

White, M.A. & Sabbioni, E. Trace element reference values in tissues from inhabitants of the European Union. X. A study of 13 elements in blood and urine of a United Kingdom population. *Sci. Total Environ*. 216:3 (1998), pp.253–270.

Whitfield, M.: The mean ocean residence time MORT concept, a rationalization. *Mar. Chem*. 8:2 (1979), pp.101–123.

Wiggering, H. & Kerth, M. (eds): Bergehalden des Steinkohlenbergbaus – Beanspruchung und Veränderung eines industriellen Ballungsraumes. In: Wiggering, H. & Thien, R. (eds): *Geologie und Ökologie im Kontext*. Vieweg, Wiesbaden, 1991, p.246.

Williams, D.R.: Earth fact sheet, 2010. Available at http://nssdc.gsfc.nasa.gov/planetary/factsheet/earth-fact.html; last access September 2010

Williams, M.: Arsenic in mine waters: an international study. *Environ. Geol.* 40:3 (2001), pp.267–278.

Winter, M.: Web elements 2.0., 2010. Available at http://www.webelements.com/arsenic/geology.html; Univ. of Sheffield.

Wood, J.M.: Biological cycles for toxic elements in the environment. *Science* 183:4129 (1974), pp.1049–1052.

Yamauchi, H., Takahashi, K., Mashiko, M., Saitoh, J. & Yamamura, Y.: Intake of different chemical species of dietary arsenic by the Japanese, and their blood and urinary arsenic levels. *Appl. Organomet. Chem.* 6:4 (1992), pp.383–388.

Ziegler, G. & Gabriel, B.: Natürliche und anthropogen überprägte Grundwasserbeschaffenheit in Festgesteinsaquiferen. In: Matschullat, J., Tobschall, H.J. & Voigt, H.J. (eds): *Geochemie und Umwelt. Relevante Prozesse in Atmo-, Pedo- und Hydrosphäre.* Springer, Berlin, 1997, pp.343–358.

CHAPTER 2

Arsenic toxicology – A review

Eduardo Mello De Capitani

2.1 TOXICOLOGY

A physiologic role of arsenic, affecting the methionine metabolism in animals, characterizes this element as essential for rats, hamsters, goats, minipigs and chicken (NRC 1999; Thornton 1999; Uthus 1990). Arsenic supplementation seems to have a growth-stimulating effect at very high doses in these animals (NRC 1999). However, arsenic does not fulfill one of the main criteria for being essential in humans, that is, *"the reduction of exposure to the element below a certain limit must result consistently and reproducibly in an impairment of physiologically important functions, and restitution of the element under otherwise identical conditions might prevent the impairment"* (NRC 1999). A controversy still exists regarding its function in human metabolism, although Mayer *et al.* (1993) suggest that arsenic deficiency might contribute to the increased risk of death in hemodialysis patients. However, arsenic has not been found to be required for any essential human biochemical process (NRC 1999). Natural As concentrations in non-polluted environments are probably sufficient to provide any supposedly nutritional need by humans (▶ 1, 13; Thornton 1999). Whether arsenic is essential was not adequately tested in humans so far, and the metabolic necessity of the element is unknown. The latter is being considered to be <0.01 mg day^{-1}, based on an average body content of 7 mg (0.1 mg As kg^{-1}) found in a regular adult of 70 kg. A non-exposed person can have at least 1 mg As kg^{-1} in the hair (Emsley 2005; ▶ 1: Table 1.6; ▶ 14).

Arsenic toxicity is directly related to the As species involved. In general, organic compounds, either the ones occurring in seafood, or the metabolites of inorganic As forms, are much less toxic than the inorganic forms. Among the inorganic species, trivalent As compounds are more toxic than pentavalent. LD$_{50}$ values (Lethal Dose for 50% of the experimental animals in acute dose testing in mg kg^{-1}) in rats is estimated to be 3 for arsine (AsH$_3$); 14 for arsenites (trivalent As); 20 for arsenates (pentavalent As); 700–1,800 for methylarsonic acid (MMA); 700–2,600 for dimethylarsinic acid (DMA); 6,500 for arsenocoline; and >10,000 for arsenobetaine (Le *et al.* 1994). All kinds of seafood contain significant amounts of organic As compounds, such as arsenobetaine and arsenosugars. Fortunately and differently from other toxic metals, arsenic can be detoxified through a series of metabolic processes present in humans and other mammalian animals.

Since ancient times arsenic is linked to a tragic history related to its use in suicides and homicides. Its popularity is due more to its low cost, to being odourless, tasteless, and its easy availability than its acute toxic efficacy compared, for instance, with cyanides. The frequency of use for homicide purposes began declining in the 19th Century, when a sensitive and practical chemical test began to be used in As detection in biological fluids and tissues (Gorby 1994). On the other hand, arsenic has been used as medicine for a great variety of ailments since Hippocrates. Hippocrates himself used to prescribe realgar (As$_2$S$_2$) and orpiment (As$_2$S$_3$) as caustic agents in the treatment of skin ulcers (Gorby 1994). It has been a popular belief, since ancient times, that arsenic in small doses can work as a tonic and energizing substance, increasing appetite, body weight and general vigour. During the 19th Century arsenic was a stem medicine, included in *Materia Medica*, as a panacea for hundreds of illnesses. Fowler's solution (1% potassium arsenite, KAsO$_2$) was the most common way of prescribing

27

arsenic at that time. In 1907, Erlich synthesized arsphenamine ($C_{12}H_{14}As_2Cl_2N_2O_2$), an organoarsenical compound registered as Salvarsan 606 and used in the treatment of syphilis until the advent of penicillin in 1943 (Gorby 1994). See below for current usage.

2.2   TOXICOKINETICS

Absorption. Arsenic, in all its chemical forms, is odourless and tasteless. Solubility in water is a key factor in absorption, and soluble As species, as trivalent and pentavalent compounds, are well absorbed within a few hours in the gastrointestinal tract of experimental animals and humans (60 to 90% – Marafante and Vahter 1987). The highest absorption occurs predominantly in the small intestine, followed by the colon (Ford 2002). Arsenic trioxide (As_2O_3), which is only slightly soluble in water, has a lower absorption rate, depending on the pH of the gastric juice and the size of the ingested particle (Vather 1983). However, As_2O_3, when ingested in aqueous solution, is better absorbed than the same amount given with food, thus increasing the risk of intoxication (Ford 2002). Experimental studies on monkeys have shown a difference of a 3 to 4 fold increase in the absorption of inorganic arsenic given as gavage (force feeding), compared to the same As salts given orally jointly with domestic dust or soil (Freeman *et al.* 1995). This is an important observation when considering human exposure to dust or soil arsenic (▶ 12), and comparable As amounts in drinking water (▶ 11). Arsenic sulfides, selenites and As-Pb compounds are insoluble and considered to not be absorbed (ATSDR 2000).

Organic arsenic compounds present in seafood, being fat soluble, are readily absorbed. The average As concentration in urine of subjects without known As exposure may vary from <<10 to 15 μg L^{-1} (▶ 14). After the intake of a single meal of fish or shellfish, arsenic in urine may increase to more than 1,000 μg L^{-1} (Vahter 1994).

Dermal absorption seems to be negligible in most As species. Dermal absorption in Rhesus monkeys has been shown to be only 3–5% after contact with As-rich soil and water (Wester *et al.* 1993). Despite the low rate, it is advisable to wash the skin with soap and clean water after this kind of exposure, even if it is a long-term low-dose exposure situation (▶ 15). Some systemic effects were seen after extensive skin contact with As acids and As trichloride, due to their corrosive action that disintegrates the skin's natural protection layer, facilitating absorption (Robinson 1975).

Respiratory absorption is lower than gastrointestinal, reaching 30%, depending mostly on solubility and particle size. Large particles tend to be deposited in the upper respiratory tract where natural ciliary removal mechanisms result in transferring these particles to the gastrointestinal tract by swallowing. Particles reaching the lower respiratory tract are more easily absorbed. The principal form of airborne arsenic in the industrial setting is As trioxide (Pinto *et al.* 1976; Offergelt *et al.* 1992).

Distribution. Arsenic is rapidly distributed through the blood after absorption by the lungs or the gastrointestinal tract. Although arsenic is uniformly distributed through all organs and tissues, it initially accumulates in the liver, kidneys and lungs. A fast clearance from these organs occurs, however, and after two to four weeks, most of the remaining arsenic in the body is bound to tissues rich in cystein-containing proteins, such as skin, hair, and nails.

In high dose acute exposures, like the ones in suicide or murder attempts, the liver and the kidneys show the major As concentration in autopsied cases, and the trivalent form is the predominant species. A deposition hierarchy of arsenic in the tissues could be provisionally established as: liver > kidneys > muscles > heart > spleen > pancreas > lungs > brain (cerebellum > brain tissue) > skin > blood (erythrocytes), and to a much lesser extent, bones and teeth (ATSDR 2000; Benramdane *et al.* 1999). However, rats accumulate a large part of the absorbed arsenic in erythrocytes, binding it to the hemoglobin after methylation to dimethylarsinic acid (DMA; Vather *et al.* 1984).

Inorganic and organic As compounds can be detected in human milk, may cross the placenta and can accumulate in the fetus (brain, liver and kidneys; ATSDR 2000; Lugo *et al.* 1969).

Biotransformation of inorganic arsenic occurs by methylation of arsenite (As^{III}) to methyl-arsonic acid (MMA) and dimethylarsinic acid (DMA), mainly in the liver. The efficiency of this mechanism decreases with increasing As dose. Methylation renders inorganic As species less reactive to tissues (consequently, less toxic), facilitating the process of elimination by kidney excretion.

Oral ingestion of MMA and DMA by human volunteers resulted in direct excretion of 75% and 78% of the dose intact in the urine, respectively, against 45% after ingestion of inorganic arsenic at a similar dose (Buchet *et al.* 1981; ▶ 14). The estimated LD_{50} for humans of As trioxide is < 5.0 mg kg^{-1}; that of MMA is 50 mg kg^{-1} and of DMA 500 mg kg^{-1}, showing that each methyl group, which is added in the process, decreases the acute toxicity by at least an order of magnitude (Ford 1994). That observation might not be true for chronic exposure regarding carcinogenic effects (▶ 9, 14). Like all other enzymatic processes, there could be a limiting As dose that saturates the methylation capacity. Unfortunately that dose cannot yet be precisely defined. A lower As methylation capacity can also be associated with higher concentrations of inorganic arsenic in tissues, increasing, for instance, its carcinogenic effect.

Another As biotransformation mechanism depends on oxidation-reduction reactions interconverting arsenate (As^V) and arsenite (As^{III}) which occurs in a non-enzymatic way using glutathione (GSH) as the reductant (Aposhian *et al.* 1999). Reduction of arsenate to arsenite can also be mediated by arsenate reductase, mainly in the liver (Aposhian *et al.* 1999). Both methylation and oxidation-reduction processes are responsible for the biotransformation and renal excretion of more than 75% of absorbed arsenic (ATSDR 2000).

The urinary excretion pattern of As species after inorganic As exposure is 10–15% inorganic arsenic; 10–15% MMA and 60–80% DMA (Vahter 1994; ▶ 14). However, this distribution can be different according to the human population studied. For instance, studies in Argentina, investigating As exposure by drinking water, have shown that the MMA excretion was on average only 2–4%. A similar rate of MMA excretion (<5%) was seen in Chile among non-European descendents (Vahter *et al.* 1995; Hopenhayn-Rich *et al.* 1996). In Taiwan, the average MMA-excretion was more than 25% (Chiou *et al.* 1997). Thus, rates of MMA excretion may vary from 2–27% according to the few published studies, indicating that there probably is an enzymatic genetic polymorphism associated with these differences, as already reported in many other human methyltransferase activities (Vahter 1999).

Environmental factors cannot be discarded as partially influencing these results. Food habits may have a large influence on this pattern. The regular ingestion of seaweed (containing one single kind of arsenosugar) for example, as seen in Japan and Japanese communities all over the world, may result in DMA-increase in urine and no change in the amount of inorganic forms. This indicates that there is no metabolic path from organoarsenicals to inorganic forms, not even to arsenate (Le *et al.* 1994). The nutritional status can also influence As toxicity and urinary pattern of metabolite excretion. Low availability of dietary methyl groups may limit the rate of As methylation, predisposing the As concentration increase in tissue. Decrease in DMA-excretion was seen in experimental studies when animals were fed with a low protein, low methionine, and low choline diet (Buchett and Lauwerys 1985; Vather 1999). Low dietary ingestion of vitamin B12 or folate (natural folic acid) increases As methylation in vitro (NRC 1999). The use of chelating agents in humans exposed to high As concentrations in drinking water seems to block the methylation of MMA to DMA, increasing the proportion of MMA in urine up to 42% (Aposhian *et al.* 1999).

Differences in methylation can also be seen with age. Children, being more susceptible to As toxicity, tend to have a lower percentage of DMA and more inorganic arsenic in urine compared to adults (Concha *et al.* 1998a; Kurttio *et al.* 1998; ▶ 14). Gender does not seem to be an important factor in this regard (Chiou *et al.* 1997; Kurttio *et al.* 1998). Pregnant women have higher relative amount of DMA compared to non-pregnant women, suggesting a hormonal influence in the biotransformation pattern (Concha *et al.* 1998b)

Excretion. Experimental studies have shown that renal excretion of unchanged inorganic arsenic and its methylated metabolites (MMA and DMA) occurs via glomerular filtration, tubular secretion and active tubular reabsorption (Tsukamoto *et al.* 1983; Vahter and Norin 1980).

Urinary As elimination in humans occurs at a rate of 45–70% during the first 4–5 days after ingestion. Thirty percent will be eliminated with a half life of more than 1 week, and the remainder will be slowly excreted with a half life greater than 1 month (Buchet *et al.* 1981; Johnson and Farmer 1991; McKinney 1992; Pomroy *et al.* 1980; Tam *et al.* 1979; ▶ 14). Fecal elimination in humans is considered negligible, varying from 0.21–6.1% in some studies (Ford 2002; Tam *et al.* 1979). Arsenobetaine is eliminated from urine unchanged and more rapidly than inorganic forms. In humans, 25% is excreted in the first 2–4 hours; 50% during the next 20 hours; and 70% after 160 hours (Johnson and Farmer 1991). During the first day, 75–85% of ingested MMA and DMA are excreted (Buchet *et al.* 1981). Concha *et al.* (1998a) observed a very low As concentration in breast milk of mothers exposed to high As levels in drinking water (average 3.1 µg kg^{-1} during 4 months post partum) with no association with As blood levels. The same authors found a good correlation between maternal blood and cord blood arsenic ($r^2 = 0.62$; $p = 0.004$; Concha *et al.* 1998a).

Toxicodynamics is the study of the relationship between toxicant concentration and effect, with specific emphasis on action mechanisms. It involves understanding physiological, biochemical and molecular effects of toxic substances (Hodgson *et al.* 1998). The basic toxic mechanisms by which arsenic produces noxious effects are related to a few known cellular mechanisms: a) inhibition of cellular respiration at the level of the mitochondrion; b) interference in the heme homeostasis and porphyrins metabolism; c) increasing expression of the major stress proteins (mostly by cell stimulation of production of reactive oxygen species, ROS). On the other hand, arsenic can induce the production of metallothionein, a protein that binds arsenic (as well as cadmium, mercury and many essential metals) and is supposed to be one of the adaptive mechanisms for tolerance to As toxicity (Aposhian *et al.* 1999; NRC 1999). The first two toxic mechanisms and the affinity to metallothionein are secondary to the high As affinity with sulfhydryl groups, favoring the formation of quite stable complexes with enzymes containing dithiol groups. Arsenic can inactivate more than 200 enzymes, most of them involved in the cellular energy pathways and DNA synthesis and repair (Ratnaike 2003). Metallothionein is a small protein with 33% cysteine (a sulfhydryl-containing amino acid) in its molecular structure (Aposhian *et al.* 1999).

Much of the knowledge and understanding of As toxic mechanisms derives from experimental studies started in the late 19th Century, from the development of organic arsenical drugs in the early 20th Century, and from scientific effort during the 1930s and 1940s to find an effective antidote to the arsenical warfare agents used during WWI (Gorby 1994). It was for the period of this latter search that it was understood that arsenicals could complex with two sulfhydryl groups inactivating the vital pyruvate oxidase system. This stimulated testing of various dithiol compounds, and led to the discovery of BAL (British anti-lewisite), or dimercaprol ($C_3H_8OS_2$) as an effective As antidote.

Despite all the empirical use and research done during the last two centuries with organic and inorganic arsenical compounds aiming to cure a great deal of human ailments like anorexia, neuralgia, rheumatism, asthma, chorea, tuberculosis, diabetes, fevers of all kinds, skin disorders, syphilis, gonorrhea, etc., its therapeutic use is nowadays limited to two situations. One is the treatment of the meningoencephalic stage of Gambian and Rhodesian trypanosomiasis with Melasorprol, an arsenoxide derivative of an organic arsenical compound. The other clinical situation is the treatment of acute promyelocytic leukemia (APL) with Trisenox, an As trioxide derivative, with remission rates ranging from 70–90% in newly diagnosed patients, and 65–>90% in relapsed patients (Ford 2002; Miller *et al.* 2002).

2.3   CLINICAL MANIFESTATIONS

Clinical manifestations of As exposure depend basically on the arsenical species involved, and on the dose and duration of exposure. Acute fatal doses of inorganic arsenic, like As trioxide, commonly used in suicide and homicide attempts in the past (although some cases are still seen nowadays), are in the range of 100–300 mg for a regular adult, although 20 mg has shown to be life-threatening. The minimum lethal dose reported has been 1 mg kg^{-1} in a child (estimated LD$_{50}$ for humans = <5.0 mg kg^{-1}; Dart 1992; Ford 2002; Gorby 1994).

Signs and symptoms after high or lethal dose ingestion occur within 30 minutes, being delayed by simultaneous ingestion of food. Due to the increased vascular permeability caused by arsenic, the clinical initial presentation indicates hypovolemia, that is loss of intravascular liquid, causing dehydration and life-threatening electrolyte imbalance. Patients can present with intense thirst, diarrhea, painful vomiting, abdominal pain (colic), blood loss, and metallic or garlic taste. Despite the availability of BAL as an effective antidote, odds can be against any therapeutic measure depending on the time of hospital arrival. Impairment of neurological functions and hepatic and renal failure rapidly follow.

Long term oral exposure can lead to multi organ manifestations classified as carcinogenic or non-carcinogenic. Both manifestations will be described together according to the targeted organ.

Dermal effects. The skin is the major organ of As accumulation, presenting multiple dermatologic lesions. Erythematous lesions can phase into hyperpigmentation (dark brown patches). Hyperpigmentation is the initial and most frequent skin alteration (melanosis). It develops in more heavily pigmented body areas (like axillae, groin and nipple areola) or areas exposed to increased pressure, like the waist (Ford 2002; Shannon and Strayer 1989). It can occur with a "raindrop" pattern. Examination of the lesions show As deposits and increased melanin levels. Fingernails may become thin and brittle presenting whitish lines (Mees' lines, also seen in chronic thallium exposure; Mees 1919).

Hyperkeratosis typically occurs on the soles and palms of feet and hands, spreading also to fingers, toes, legs, arms and the dorsum of the hands. It is a lesion of late onset, with average latency time of 23 to 28 years (Ford 2002; Haque *et al.* 2003; Wong *et al.* 1998). Hyperkeratosis occurs in the form of small nodules, which may coalesce forming plaques as verrucous growths and diffuse keratosis. Bowen's disease is the name given to in situ squamous cell carcinoma, occurring on the trunk in multiple lesions varying from 1–10 mm. Concomitant with Bowen's disease, real squamous cell carcinoma can develop from Bowen's lesions (which are always of high malignancy grade and more aggressive, giving metastasis) or from keratotic lesions. Arsenic-induced squamous cell carcinomas are frequently multiple and more common in the extremities (palms and soles). Arsenic-induced basal cell carcinomas (a skin cancer with very low grade of malignancy) are almost always multiple, occurring frequently in the trunk. Hyperpigmentation and keratosis may regress or persist indefinitely even after the discontinuation of the exposure (Shannon and Strayer 1989). All kinds of As-induced skin lesions showed a marked dose-response relationship in all epidemiologic studies (Haque *et al.* 2003). The youngest subjects to present As-induced skin alterations were 2 and 3 years old, respectively, in Chile and China (NRC 1999).

Neurological effects. Delayed onset of peripheral or central nervous system involvement can occur between 1–5 days after acute ingestion of high doses of inorganic arsenic. The clinical picture varies from mental confusion, headaches, to severe encephalopathy with seizures and coma. Peripheral neuropathy with a sensorimotor axonopathy pattern may start within 1–4 weeks and may last for as long as two years (Ratnaike 2003). Subacute and chronic As exposures from drinking water show correlation to peripheral neuropathy diagnosis only in situations of exposure to high doses (>10 mg L^{-1}; NRC 1999). In the cohorts of people affected with arsenical-induced skin diseases in Taiwan, Argentina and Chile, researchers were not able to find a significant correlation between peripheral neuropathy and As exposure (NRC 1999). Although organic As compounds have been considered less toxic, a recent case of suicide attempt with high-dose ingestion of MSMA (monosodium methyl arsenate acid – used as herbicide) resulted in neurological sequel with bilateral deafness and peripheral sensorimotor neuropathy (De Capitani *et al.* 2005).

Cardiovascular effects. Cardiovascular instability (presented as sinus tachycardia, ventricular arrhythmias, severe hypotension, myocardial dysfunction and shock) can occur after ingestion of high doses of As trioxide contributing to the death of most patients (Ford 2002; Ratnaike 2003).

Chronic ingestion of inorganic arsenic from drinking water has been associated with the development of hypertension and ischemic heart disease (Hsueh *et al.* 1998; Rahman *et al.* 1999; Tsai *et al.* 1999). An endemic peripheral vascular disease, named Black Foot Disease (BFD), confined to the southwestern coast of the island, has been observed in Taiwan, since the beginning of the last century, with peak incidence between 1956 and 1960. Prevalence rates range from 6.51–18.85 per 1,000 inhabitants. The disease is characterized by a progressive arterial occlusion of the lower extremities, and in rare cases, of the upper extremities leading to ulcerations, gangrene and spontaneous or surgical amputations. BFD usually occurs together with arsenic skin alterations. Histopathological examination of some patients showed the presence of thromboangiitis and arteriosclerosis obliterans as the basic lesions. The incidence of BFD markedly decreased after the distribution of As free tap water to the affected communities in the past 2–3 decades (NRC 1999; Tseng 2002). Different symptoms related to peripheral vascular disease, not as dramatic as BFD, were also observed in Antofagasta, Chile, epidemiologically associated with arsenic in drinking water (NRC 1999). Cerebrovascular disease has also been associated to As ingestion in the same geographic areas (NRC 1999).

Hematological effects. Anemia, leucopenia and thrombocytopenia can be expected among people chronically exposed to inorganic arsenic. As stressed before, the inhibition of almost 200 enzymes in the regular human metabolism can also render the haemopoietic system susceptible to toxic As action. The anemia may also be the result of hemolysis and also the action on the erythrocyte production. More severe cases can present aplastic anemia progressing eventually to acute leukemia (Gorby 1994). All the hematological alterations, except for leukemia, can regress after the end of exposure.

Gastrointestinal effects. Acute high-dose ingestion of inorganic arsenic can result in nausea, vomiting, diarrhea and abdominal pain, and gastrointestinal hemorrhage. This is due to a direct irritating action. It can also be seen to a lesser degree, in long-term consumption of drinking water above an exposure of 0.005 mg As kg^{-1} day^{-1} (ASTDR 2000). Organic salts, like MSMA (monosodium methyl arsenate acid) can provoke vomits and cause abdominal pain (De Capitani *et al.* 2005; Shum *et al.* 1995).

Hepatic effects. Hepatic injury, from alteration in blood hepatic enzyme levels to hepatic fibrosis, has been described in acute and long-term exposure to inorganic arsenicals (ATSDR 2000). A study in India reported hepatomegaly in 76% of patients with chronic arsenicism, with liver biopsy showing noncirrhotic portal fibrosis in 91% (Santra *et al.* 1999). It seems that these effects can be observed in chronic exposures to doses ranging from 0.01–0.1 mg kg^{-1} day^{-1} (ASTDR 2000). Hepatic angiosarcomas have been linked to chronic As exposure. Evidence was first published in 1942 after the investigation of German workers exposed to As pesticides at the beginning of the 20th Century (Morton and Dunnete 1994). Hepatic enzymes and bilirubin elevation were reported after the ingestion of 1,714 mg MSMA kg^{-1}, which returned to normal levels after chelating therapy (De Capitani *et al.* 2005).

Renal effects. During acute episodes of As intoxication, renal failure can occur mostly due to the loss of fluid and hypovolemic shock, resulting from vascular permeability alteration, or less frequently, due to hemolysis. Studies in experimental animals show that the kidney is not a major target organ for inorganic or organic arsenic, although it is the major route of As excretion, as well as a major site of conversion of As$^{(V)}$ to the trivalent form (ASTDR 2000). Long-term As exposures have not shown overt renal alterations in the populations studied, suggesting that the kidney is less sensitive to toxic As effects than other organs. Garllip *et al.* (2003), studying the protein excretion pattern in a population chronically exposed to low As doses, showed a statistical correlation of urinary As levels with β2-microglobulin excretion (marker of early glomerular lesion). These results, however, must be confirmed in similar exposed populations.

Respiratory effects. Lungs and airways can be affected by inhalation of As fumes and dust mainly in workplaces or pesticide-spraying activities. Rhinitis, laryngitis, pharyngitis, tracheobronchitis and asthma may occur secondary to the irritating As properties (Morton and Dunette 1994). No such problems were diagnosed in populations chronically exposed to arsenic by ingestion. However, the most prominent As effect on the respiratory system is the high prevalence of lung cancer after long-term exposure by inhalation, and in some communities, by ingestion. This has been confirmed by many epidemiological studies of exposed workers or of communities living around copper smelters (Enterline and Marsh 1982; Enterline *et al.* 1987; Järup and Pershagen 1991). Chen *et al.* (2004) showed a significant dose-response relationship of ingested arsenic on lung cancer risk, with an evident synergistic effect with cigarette smoking. Arsenic ingestion in Argentina was correlated with high risk for lung and renal cancer and not to liver and skin cancer (Hopenhayn-Rich *et al.* 1998).

Reproductive effects. As discussed above, arsenic can cross the placental barrier concentrating on the fetus. Congenital malformations are well documented in many species of experimental animals and show that the nature of the malformations depend on the timing of the exposure (Morton and Dunette 1994). There are few studies on human reproductive outcomes associated with As exposure. Some studies showed a higher rate of spontaneous abortion and low weight babies from mothers working in or living near copper smelters after inhalatory As exposure (NRC 1999). The presence of many other toxic contaminants in the workplace might act as confounding factors complicating the task of establishing a real causal relationship for the arsenic alone. An epidemiological study in Chile (Antofagasta) showed an elevation of the late fetal, neonatal, and postneonatal mortality rates from exposed mothers compared to a community (Valparaiso) with low As concentration in water (Hopenhayn-Rich *et al.* 2000).

Effects on other organs and systems. Increased prevalence of diabetes mellitus in populations chronically As exposed from drinking water, indicates the existence of a toxic endocrinological effect. This effect is probably secondary to inhibition of the pyruvate dihydrogenase system responsible for the normal carbohydrate metabolism. Experimental studies have already shown a glucose intolerance effect (NRC 1999). Tseng *et al.* (2002) reviewed the epidemiologic studies reporting the association between chronic inhaled arsenic (glass workers in Sweden) and ingested arsenic (communities in Taiwan and Bangladesh) and development of diabetes mellitus, concluding that, despite the consistent findings supporting this association, some methodological limitations might reduce the strength of the association. Nevertheless, a better explanation of biochemical toxicity must still be elucidated for this effect. There is no convincing epidemiological or experimental data suggesting that arsenic is an endocrine disruptor (ASTDR 2000).

Carcinogenic effects. More than a hundred years ago, Hutchinson (1887, in Sommers and McManus 1953) observed a high prevalence of skin tumors in people chronically medicated with arsenicals (Fowler's solution and many other preparations available at the time). In 1973, the International Agency for Research on Cancer (IARC), a World Health Organization body, first evaluated arsenic for its carcinogenic effects, and found overwhelming evidence based on epidemiological studies, to consider the existence of a causal relationship between skin cancer and exposure to high doses of inorganic arsenic from water ingestion or from inhalation at work. The same relationship was not accepted for cancer in other organs at the time. Part of this cautious position was due to the absence of experimental evidence of a causative role for As carcinogenicity (IARC 1973).

In a reevaluation process in 1979, IARC definitely classified certain As compounds in Group I (only for the skin). Arsenic was thus considered an element definitely carcinogenic to humans, notwithstanding the lack of experimental animal evidence. Soon after, IARC published a statement considering arsenic a cause of lung cancer in humans (IARC 1980). In 1987, in a supplementary publication, the IARC committee considered arsenic carcinogenic to humans (Group 1), on the basis of sufficient evidence for an increased risk for skin cancer among patients exposed to inorganic arsenic through medical treatment and an increased risk for lung cancer among workers involved in mining and smelting, who inhaled inorganic arsenic. Limited evidence of carcinogenicity in experimental animals was still a problem (IARC 1987).

In 2004, IARC published a monograph containing a re-evaluation of the evidence for As carcinogenicity in humans and animals. It concludes that there is: 1) sufficient evidence in humans that arsenic in drinking-water causes cancers of the urinary bladder, lung and skin; 2) sufficient evidence in experimental animals for the carcinogenicity of dimethylarsinic acid (DMA); 3) limited evidence in experimental animals for the carcinogenicity of sodium arsenite, calcium arsenate and As trioxide; and 4) inadequate evidence in experimental animals for the carcinogenicity of sodium arsenate and arsenic trisulfide. Taken together, IARC says that the studies on inorganic arsenic still provide limited evidence for carcinogenicity in experimental animals (IARC 2004).

The United States National Research Council (NRC) concluded in a 1999 report that *"there was sufficient evidence from human epidemiological studies in Taiwan, Chile, and Argentina that chronic ingestion of inorganic arsenic causes bladder and lung cancer, as well as skin cancer"* (NRC 1999). This conclusion was based on the evaluation of epidemiological studies in which As concentrations in drinking water were at least several hundred micrograms per liter, and that very few studies had addressed the risk of cancer at lower doses (NRC 1999). However, in its update report (NRC 2001), the committee found that people who daily consume water with 3 μg As L^{-1} have about a 1 : 1,000 increased risk of developing bladder or lung cancer during their lifetime. This risk increases with concentration: with 5 μg As L^{-1} (1.5 : 1,000), 10 μg As L^{-1} (>3: 1,000), and with 20 μg As L^{-1} (close to 7 : 1,000).

Carcinogenicity in other human organs (liver, kidneys and haemopoietic system) supposedly exists. Weak evidence is based on reports of cases, series of cases, or isolated epidemiological studies.

REFERENCES

Aposhian, H.V., Zakharyan, R.A., Wildfang, E.K., Healy, S.M., Gailer, J., Radabaugh, R., Bogdan, G.M., Powell, L.A. & Aposhian, M.M.: How is inorganic arsenic detoxified? In: Chappell, W.R. Abernathy, C.O. & Calderon, R.I.: (eds): *Arsenic exposure and health effects*, Volume III. Elsevier, Amsterdam, 1999, pp.289–297.

ATSDR: Toxicological profile for arsenic. US Department of Health and Human Services, Agency for Toxic Substances and Disease Registry, Atlanta, USA, 2000. For latest updates, see: http://www.atsdr.cdc.gov/toxprofiles/tp.asp?id=22&tid=3.

Benramdane, L., Accominotti, M., Fanton, L., Malicier, D. & Vallon, J.J.: Arsenic speciation in human organs following fatal arsenic trioxide poisoning – a case report. *Clin. Chem.* 45:2 (1999), pp.301–306.

Buchet, J.P. & Lauwerys, R.: Study of inorganic arsenic methylation by rat liver in vitro: relevance for the interpretation of observations in man. *Arch. Toxicol.* 57:2 (1985), pp.125–129.

Buchet, J.P., Lauwerys, R. & Roels, H.: Comparison of urinary excretion of arsenic metabolites after a single dose of sodium arsenite, monomethylarsonate or dimethylarsinate in man. *Int. Arch. Occup. Environ. Health* 48:1 (1981), pp.71–79.

Chen, C.L., Hsu, L.I., Chiou, H.Y., Hsueh, Y.M., Chen, S.Y., Wu, M.M. & Chen, C.J.: Ingested arsenic, cigarette smoking, and lung cancer risk: a follow-up study in arseniasis-endemic areas in Taiwan. *JAMA* 292:24 (2004), pp.2984–2990.

Chiou, H.Y., Hsueh, Y.M., Hsieh, L.L., Hsu, L.I., Hsu, Y.H., Hsieh, F.I., Wei, M.L., Chen, H.C., Yang, H.T., Leu, L.C., Chu, T.H., Chen-Wu, C., Yang, M.H. & Chen, C.J.: Arsenic methylation capacity, body retention, and null genotypes of glutathione S-transferase M1 and T1 among current arsenic-exposed residents in Taiwan. *Mutat. Res.* 386:3 (1997), pp.197–207.

Concha, G., Vogler, G., Lezcano, D., Nermell, B. & Vahter, M.: Exposure to inorganic arsenic metabolites during early human development. *Toxicol. Sci.* 44:2 (1998a), pp.185–190.

Concha, G., Nermell, B. & Vahter, M.: Metabolism of inorganic arsenic in children with chronic high arsenic exposure in northern Argentina. *Environ. Health Perspect.* 106:6 (1998b), pp.355–359.

Dart, R.C.: Arsenic. In: Sullivan, J.B. & Krieger, G.R. (eds): *Hazardous materials toxicology – Clinical principles of environmental health*. Williams and Wilkins, Baltimore, 1992, pp.818–823.

De Capitani, E.M., Vieira, R.J., Madureira, P.R., Mello, S.M., Kira, C.S., Soubhia, P.C. & Toledo, A.S.: Auditory neurotoxicity and hepatotoxicity after MSMA (monosodium methanarsenate) high dose oral intake. *Clin. Toxicol.* 43:4 (2005), pp.287–289.

Emsley, J.: *Elements of murder*. Oxford University Press, 2005, p.460.

Enterline, P.E. & Marsh, G.M.: Cancer among workers exposed to arsenic and other substances in a copper smelter. *Am. J. Epidemiol.* 116:6 (1982), pp.895–911.

Enterline, P.E., Henderson, V.L. & Marsh, G.M.: Exposure to arsenic and respiratory cancer. A reanalysis. *Am. J. Epidemiol.* 125:6 (1987), pp.929–938.

Ford, M.: Arsenic. In: Goldfrank, L.R., Flomenbaum, N.E., Lewin, N.A., Howland, M.A., Hoffman, R.S. & Nelson, L.S. (eds): *Goldfrank's Toxicology Emergencies*, 5th ed, Appleton & Lange, Norwalk, 1994, pp.1011–1025.

Ford, M.: Arsenic. In: Goldfrank, L.R., Flomenbaum, N.E., Lewin, N.A., Howland, M.A., Hoffman, R.S. & Nelson, L.S. (eds): *Goldfrank's Toxicology Emergencies* 7th ed, McGraw-Hill, New York, 2002, pp.1183–1195.

Freeman, G.B., Schoof, R.A., Ruby, M.V., Davis, A.O., Dill, J.A., Liao, S.C., Lapin, C.A. & Bergstrom, P.D.: Bioavailability of arsenic in soil and house dust impacted by smelter activities following oral administration in cynomolgus monkeys. *Fundam. Appl. Toxicol.* 28:2 (1995), pp.215–222.

Garllip, C.R., Bottini, P.V., De Capitani, Pinho, M.C., Panzan, A.N.D., Sakuma, A.M.A. & Paoliello, M.B.: Urinary protein excretion profile: a contribution for subclinical renal damage identification among environmental heavy metals exposure in Southeast Brazil. *Internat Conf. Heavy Metal Environ.* 12, Grenoble, France J. *Phys. IV France* 107:1 (2003), pp.513–516.

Gorby, M.S.: Arsenic in human medicine. In: Niriagu, J.O. (ed): *Arsenic in the environment*, Volume II. John Wiley & Sons, New York, 1994, pp.1–16.

Haque, R., Mazumder, D.N., Samanta, S., Ghosh, N., Kalman, D., Smith, M.M., Mitra, S., Santra, A., Lahiri, S., Das, S., De, B.K. & Smith, A.H.: Arsenic in drinking water and skin lesions: dose-response data from West Bengal, India. *Epidemiol.* 14:2 (2003), pp.174–182.

Hodgson, E., Chambers, J.E. & Mailman, R.B. (eds): *Dictionary of toxicology*. 2nd ed, Macmillan Reference Ltd/Grove's Dictionaries Inc., London, New York, 1998, p.504

Hopenhayn-Rich, C., Biggs, M.L., Kalman, D.A., Moore, L.E. & Smith, A.H.: Arsenic methylation patterns before and after changing from high to lower concentrations of arsenic in drinking water. *Environ. Health Perspect.* 104:11 (1996), pp.1200–1207.

Hopenhayn-Rich, C., Biggs, M.L. & Smith, A.H.: Lung and kidney cancer mortality associated with arsenic in drinking water in Córdoba, Argentina. *Int. J. Epidemiol.* 27:4 (1998), pp.561–569.

Hopenhayn-Rich, C., Browning, S.R., Hertz-Picciotto, I., Ferreccio, C., Peralta, C. & Gibb, H.: Chronic arsenic exposure and risk of infant mortality in two areas of Chile. *Environ. Health Perspect.* 108:7 (2000), pp.667–673.

Hsueh, Y.M., Wu, W.L., Huang, Y.L., Chiou, H.Y., Tseng, C.H. & Chen, C.J.: Low serum carotene level and increased risk of ischemic heart disease related to long-term arsenic exposure. *Atherosclerosis* 141:2 (1998), pp.249–257.

IARC: Monographs on the evaluation of carcinogenic risks to humans. 2. Some inorganic and organometallic compounds – arsenic and inorganic arsenic compounds. International Agency for Research on Cancer, 1973, p.181.

IARC: Monographs on the evaluation of carcinogenic risks to humans. 23. Arsenic and arsenic compounds. International Agency for Research on Cancer, 1980, p.438.

IARC: Overall evaluations of carcinogenicity: an updating of IARC monographs. 1–42: Supplement 7, Arsenic and arsenic compounds (Group 1). International Agency for Research on Cancer, 1987, p.440.

IARC: Monographs on the evaluation of carcinogenic risks to humans. 84. Some drinking-water disinfectants and contaminants, including arsenic. International Agency for Research on Cancer, 2004, p.512.

Järup, L. & Pershagen, G.: Arsenic exposure, smoking, and lung cancer in smelter workers – a case-control study. *Am. J. Epidemiol.* 134:6 (1991), pp.545–551.

Johnson, L.R. & Farmer, J.G.: Use of human metabolic studies and urinary arsenic speciation in assessing arsenic exposure. *Bull. Environ. Contam. Toxicol.* 46:1 (1991), pp.53–61.

Kurttio, P., Komulainen, H., Hakala, E., Kahelin, H. & Pekkanen, J.: Urinary excretion of arsenic species after exposure to arsenic present in drinking water. *Arch. Environ. Contam. Toxicol.* 34:3 (1998), pp.297–305.

Le, X.C., Cullen, W.R. & Reimer, K.J.: Human urinary arsenic excretion after one-time ingestion of seaweed, crab, and shrimp. *Clin. Chem.* 40:4: (1994), pp.617–624.

Lugo, G., Cassady, G. & Palmisano, P.: Acute maternal arsenic intoxication with neonatal death. *Am. J. Dis. Child.* 117:3 (1969), pp.328–330.

Ma, H.Z., Xia, Y.J., Wu, K.G., Sun, T.Z. & Mumford, J.L.: Human exposure to arsenic and health effects in Bayingnormen, Inner Mongolia. In: Chappell, W.R., Abernathy, C.O. & Calderon, R.I. (eds): *Arsenic exposure and health effects*, Volume III. Elsevier, Amsterdam, 1999, pp.127–131.

Marafante, E. & Vahter, M.: Solubility, retention, and metabolism of intratracheally and orally administered inorganic arsenic compounds in the hamster. *Environ. Res.* 42:1 (1987), pp.72–86.

Mayer, D.R., Kosmus, W., Pogglitsch, H., Mayer, D. & Beyer, W.: Essential trace elements in humans. Serum arsenic concentrations in hemodialysis patients in comparison to healthy controls. *Biol. Trace Elem. Res.* 37:1 pp.27–38.

McKinney, J.D.: Metabolism and disposition of inorganic arsenic in laboratory animals and humans. *Environ. Geochem. Health* 14:2 (1992), pp.43–48.

Mees, R.A.: The nails with arsenical polyneuritis. *JAMA* 72 (1919), p.1337.

Miller, W.H. Jr., Schipper, H.M., Lee, J.S., Singer, J. & Waxman, S.: Mechanisms of action of arsenic trioxide. *Cancer Res.* 62:14 (2002), pp.3893–3903.

Morton, W.E. & Dunnete, D.A.: Health effects of environmental arsenic. In: Nriagu, J.O. (ed): *Arsenic in the environment*, Volume II. John Wiley & Sons, Inc., New York, 1994, pp.17–34.

NRC: *Arsenic in drinking water*. National Research Council; National Academy Press, Washington, DC, 1999, pp.310.

NRC: Update report on arsenic in drinking water. National Research Council; National Academy Press, Washington, DC, 2001, pp.310. http://www.nap.edu/openbook.php?isbn=0309063337&page=R1

Offergelt, J.A., Roels, H., Buchet, J.P., Boeckx, M. & Lauwerys, R.: Relation between airbone arsenic and urinary excretion of inorganic arsenic and its methylated metabolites. *Br. J. In. Med.* 49:6 (1992), pp.387–393.

Pinto, S.S., Varner, M.O., Nelson, K.W., Labbe, A.L. & White. L.D.: Arsenic trioxide absorption and excretion in industry. *J. Occup. Med.* 18:10 (1976), pp.677–680.

Pomroy, C., Charbonneau, S.M., McCullough, R.S. & Tam, G.K.: Human retention studies with [74]As. *Toxicol. Appl. Pharmacol.* 53:3 (1980), pp.550–556.

Rahman, M., Tondel, M., Ahmad, S.A., Chowdhury, I.A., Faruquee, M.H. & Axelson, O.: Hypertension and arsenic exposure in Bangladesh. *Hypertension* 33:1 (1999), pp.74–78.

Ratnaike, R.N.: Acute and chronic arsenic toxicity. *Postgrad. Med. J.* 79:933 (2003), pp.391–396.

Robinson, T.J.: Arsenical polyneuropathy due to caustic arsenic paste. *Brit. Med. J.* 3 (1975), p.139.

Santra, A., Das, G.J., De, B.K., Roy, B. & Guha Mazumder, D.N.: Hepatic manifestations in chronic arsenic toxicity. *Indian J. Gastroenterol.* 18:4 (1999), pp.152–155.

Shannon, R.L. & Strayer, D.S.: Arsenic-induced skin toxicity. *Human Toxicol.* 8:2 (1989), pp.99–104.

Shum, S., Whitehead, J., Vaughn, L., Shum, S. & Hale, T.: Chelation of organoarsenate with dimercaptosuccinic acid. *Vet. Hum. Toxicol.* 37:3 (1995), pp.239–242.

Sommers, C.S. & McManus, R.G.: Multiple arsenical cancers of skin and internal organs. *Cancer* 6:2 (1953), pp.347–359.

Tam, G.K., Charbonneau, S.M., Bryce, F., Pomroy, C. & Sandi, E.: Metabolism of inorganic arsenic ([74]As) in humans following oral ingestion. *Toxicol. Appl. Pharmacol.* 50:2 (1979), pp.319–322.

Thornton, I.: Arsenic in the global environment: looking towards the millennium. In: Chappell, W.R., Abernathy, C.O. & Calderon, R.I. (eds): *Arsenic exposure and health effects*, Volume III. Elsevier, Amsterdam, 1999, pp.1–7.

Tsai, S.M., Wang, T.N. & Ko, Y.C.: Mortality for certain diseases in areas with high levels of arsenic in drinking water. *Arch. Environ. Health* 54:3 (1999), pp.186–193.

Tseng, C.H.: An overview on peripheral vascular disease in blackfoot disease-hyperendemic villages in Taiwan. *Angiol.* 53:5 (2002), pp.529–537.

Tseng, C.H., Tseng, C.P., Chiou, H.Y., Hsueh, Y.M., Chong, C.K. & Chen, C.J.: Epidemiologic evidence of diabetogenic effect of arsenic. *Toxicol. Letters* 1331 (2002), pp.69–76.

Tsukamoto, H., Parker, H.R., Gribble, D.H., Mariassy, A. & Peoples, S.A.: Nephrotoxicity of sodium arsenate in dogs. *Am. J. Vet. Res.* 44:12 (1983), pp.2324–2330.

Uthus, E.O.: Effects of arsenic deprivation in hamsters. *Magnes. Trace Elem.* 9:4 (1990), pp.227–232.

Vahter, M.: Metabolism of arsenic. In: Fowler, B.A. (ed): *Biological and environmental effects of arsenic*. Elsevier, New York, 1983, pp.171–198.

Vahter, M.: What are the chemical forms of arsenic in urine, and what can they tell us about exposure? *Clin. Chem.* 40:5 (1994), pp.679–680.

Vahter, M.: Variation in human metabolism of arsenic. In: Chappell, W.R., Abernathy, C.O. & Calderon, R.I. (eds): *Arsenic exposure and health effects*, Volume III. Elsevier, Amsterdam, 1999, pp.267–279.

Vahter, M. & Norin, H.: Metabolism of [74]As-labeled trivalent and pentavalent inorganic arsenic in mice. *Environ. Res.* 21:2 (1980), pp.446–457.

Vahter, M., Marafante, E. & Dencker, L.: Tissue distribution and retention of [74]As-dimethylarsinic acid in mice and rats. *Arch. Environ. Contam. Toxicol.* 13:3 (1984), pp.259–264.

Vahter, M., Concha, G., Nermell, B., Nilsson, R., Dulout, F. & Natarajan, A.T.: A unique metabolism of inorganic arsenic in native Andean women. *Eur. J. Pharmacol.* 293:4 (1995), pp.455–462.

Wester, R.C., Maibach, H.I., Sedik, L., Melendres, J. & Wade, M.: In vivo and in vitro percutaneous absorption and skin decontamination of arsenic from water and soil. *Fundam. Appl. Toxicol.* 20:3 (1993), pp.336–340.

Wong, S.S., Tan, K.C. & Goh, C.L.: Cutaneous manifestations of chronic arsenicism: review of seventeen cases. *J. Am. Acad. Dermatol.* 38:2/1 (1998), pp.179–185.

CHAPTER 3

Arsenic removal from water

Wolfgang H. Höll (†) & Eleonora Deschamps

3.1 INTRODUCTION

Arsenic occurs in the natural environment in the oxidation states −3, 0, +3 and +5 and is found in about 200 different minerals including elemental arsenic, arsenides, sulphides, oxides, arsenates and arsenites (Anonymous 2000; Riedel 1994; ► 1). The most abundant mineral is arsenopyrite, FeAsS, which is often the host for gold. Arsenic is a well-known poison and is not an element essential for the human body (Anonymous 2000; ► 2). In water, arsenic is found in the two oxidation states +3 and +5 (► 11). The trivalent form, As$^{(III)}$, is hydrolysed as arsenous acid H_3AsO_3 and is present as the free acid or as one of the species resulting from its dissociation. The pentavalent form, As$^{(V)}$, is also hydrolysed as arsenic acid H_3AsO_4, and occurs as the non-dissociated acid or as their dissociated species. However, the two acids show completely different dissociation patterns (Figure 3.1). Arsenic acid is almost completely dissociated at pH values >4, while arsenous acid shows substantial dissociation only at pH values >8.

Arsenic concentrations in water bodies may range from <1 µg L^{-1} in uninfluenced surface waters to >400 µg L^{-1} in rivers and lakes affected by geothermal or industrial waste waters. Aquifer concentrations vary from very low to several mg L^{-1} from both natural and industrial sources (Anonymous 2000; ► 1). The different methods of As species elimination are based on the chemical and physical properties of the two sets of As species and their reactions with other dissolved species or with solid surfaces.

3.2 ARSENIC REMOVAL BY PRECIPITATION

The insolubility of certain inorganic As$^{(V)}$-compounds is the basis of many hydrometallurgical As removal processes. The most common methods to remove arsenic from process streams

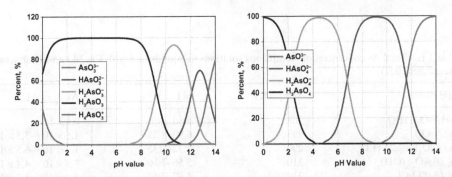

Figure 3.1. Speciation of As$^{(III)}$, (left), and As$^{(V)}$, (right). Total As concentration: 100 µg L^{-1}. The species in the legend are displayed from right to left in each figure, starting with AsO$_3^{3-}$ on the right hand side of the left figure, and with AsO$_4^{3-}$ on the right hand side of the right figure.

are by precipitation as $As^{(III)}$-sulphide, calcium arsenate or ferric arsenate. Unfortunately, all of these materials are unstable under certain conditions and therefore not suitable for direct disposal at uncontained sites as they will produce As-bearing leachates (Bothe and Brown 1999; Nishimura *et al.* 1993; Robins 2001; Robins *et al.* 2001; Twidwell *et al.* 1999).

Arsenic sulphide, As_2S_3, can be generated by adding ferrous sulphate solutions and by means of sulphate-reducing bacteria (Tenny 2001):

$$8Fe^{2+} + SO_4^{2-} + 20H_2O \Leftrightarrow 8Fe(OH)_3 + 14H^+ + H_2S$$

Arsenic is then precipitated by the sulphide:

$$2H_3AsO_4 + 5HS^- \Leftrightarrow As_2S_5 + 3H_2O + 5OH^-$$

Arsenic sulphide has its lowest solubility below pH 4, but its solubility is, in general, significantly higher than accepted by standards (▶ 5). Precipitation by direct application of hydrogen sulphide gas is not as effective and requires pH ranges of 2.5–3.0. Calcium arsenate compounds are generated by adding CaO or $Ca(OH)_2$ to contaminated waters. Different precipitates can be generated (Table 3.1). When operated at pH values >10.5, a high percentage of arsenic can be precipitated from solutions containing >50 mg As L^{-1}. It is difficult, however, to achieve final concentrations <1 mg L^{-1}, although concentrations of about 10 µg L^{-1} have been reported (Robins 2001; Robins *et al.* 2001). Solid calcium arsenate reacts with carbon dioxide to form $CaCO_3$ while arsenic is remobilised. Addition of magnesium salts leads to the formation of $Mg_3(AsO_4)_2$. To a limited extent, this method is applied to achieve arsenate fixation in soils, sediments and wastes (Magalhães 2002). $As^{(V)}$ can be removed through precipitation of ferric arsenate. One way of doing this is the addition of ferric salts to As-bearing water (Nishimura *et al.* 1993):

$$Fe^{3+} + AsO_4^{3-} \Leftrightarrow FeAsO_4(s)$$

Precipitation is possible at pH values < about pH 2 and leads to an amorphous material with particle sizes of about 100 nm. Conversion to crystalline material (scorodite) requires temperatures >90°C (Robins *et al.* 2001). Another option uses ferrous salts (e.g., ferrous sulphate) and subsequent oxidation by means of ferrate ions (Vogels and Johnson 1998):

$$Fe^{2+} + AsO_4^{3+} \Leftrightarrow FeAsO_4^-$$

$$FeAsO_4^- + FeO_4^{2-} \Leftrightarrow FeAsO_4(s) + FeO(OH)(s)$$

Table 3.1. Total As concentration in aqueous solutions in equilibrium with Ca, Mg and Fe arsenates (Magalhães 2002).

Solid phase	Temp. (K)	pH	As_{total} mol L^{-1}
$CaHAsO_4 \cdot H_2O$	308	acid	0.12–1.2
$Ca_3(AsO_4)_2$	293	6.90–8.35	1.5×10^{-2}–3.5×10^{-3}
$Ca_3(AsO_4)_2 \cdot 4.25 H_2O$	296	7.32–7.35	1.1×10^{-2}–6.5×10^{-3}
$Ca_{10}(As(O_4)_6(OH)_2$	310	5.56–7.16	7.5×10^{-3}–4.4×10^{-4}
$Ca_{10}(AsO_4)_6Cl_2$	310	4.67–7.42	1.9×10^{-3}–3.7×10^{-5}
$Mg_3(AsO_4)_2$	293	6.50–7.40	1.5×10^{-2}–4.6×10^{-3}
$FeAsO_4$	293	1.90–2.95	3.7×10^{-3}–8.5×10^{-5}
$FeAsO_4 \cdot 2H_2O$	293	5.53–6.35	1.4×10^{-4}–2.5×10^{-3}

The solubility decreases with increasing Fe_3^+-dosage (West General 2006). Ferric arsenate in either form is not thermodynamically stable in the neutral to higher pH range. The materials are also not stable in alkaline cement cast admixes (Magalhães 2002). The solubility of different precipitates is summarised in Table 3.1.

3.3   ARSENIC REMOVAL THROUGH SORPTION

Dissociation of both arsenous and arsenic acid leads to anionic species which can be sorbed onto solid surfaces which bear a positive charge. Such solids may be synthetic polymeric ion exchangers or inorganic materials, predominantly hydrous metal oxides. The latter ones can be applied as granular packed beds or as freshly precipitated hydroxides.

3.3.1   *Arsenic removal using synthetic ion exchangers*

One way of doing this is the application of synthetic ion exchange resins. These resins are usually based on a cross-linked polymer matrix (polystyrene) to which charged functional groups (various amine or quaternary ammonium) have been attached. Commercially available, strongly basic ion exchange resins in the chloride form have been proposed for As-species removal (Anonymous 2000):

$$\overline{R - [N(CH_3)_3]^+\, Cl^-} + H_2AsO_4^- \quad \Leftrightarrow \quad \overline{R - [N(CH_3)_3]^+\, H_2AsO_4^-} + Cl^-$$

(R = matrix, overbarred symbols refer to the exchanger phase). Regeneration of the resins is achieved by means of NaCl solutions. The effective exchange capacity of these exchange resins strongly depends on the composition of the raw water and the influence of competitive sorption of other anions from the background composition. The relative sorption follows the so-called selectivity series. For strongly basic resins of type 1 and common anions, the following series has been found (Clifford 1999):

$$SO_4^{2-} \gg NO_3^- > HAsO_4^{2-} > Cl^-$$

As a consequence, arsenate species will be eliminated. Unfortunately, there is a strong interference by sulphate and possibly also by nitrate anions. Efficient arsenate species elimination becomes possible only for sulphate concentrations below 50 mg L^{-1} and filter throughputs of <750 bed volumes between two regenerations can be achieved. The service cycles become too short for an economic elimination at higher concentrations. Oxidation is required prior to the ion exchange step with As[(III)] species in the raw water. Some studies report on the application of iron oxide-loaded or MnO_2-loaded anion exchangers (Lenoble *et al.* 2004).

3.3.2   *Arsenic removal using hydrous oxides/hydroxides*

The second way of As species sorption consists of the application of hydrous metal oxides among which activated alumina, and ferric oxides/hydroxides have obtained importance. These surfaces possess hydroxyl groups which are subject to protolytic reactions in contact with water. The degree of protonation or deprotonation depends on the surface properties and is characterised by the so-called point of zero charge pH value, pH_{PZC}. At pH values < pH_{PZC} the surface is predominantly protonated and bears a net positive charge. For pH > pH_{PZC}, however, the surface is predominantly deprotonated and bears a negative charge, which has to be balanced by cations. Using Me as the symbol for a trivalent metal atom at the surface, the respective reactions can be written as (Horst and Höll 1997):

$$> Me - OH + H^+ \quad \Leftrightarrow \quad > Me - OH_2^+$$

$$> Me - OH_2^+ + An^- \quad \Leftrightarrow \quad > Me - OH_2An$$

$$> Me - OH - H^+ \quad \Leftrightarrow \quad > Me - O^-$$

$$> Me - O^- + Cat^+ \quad \Leftrightarrow \quad > Me - OCat$$

The compounds generated at the surface are designated as surface complexes. Depending on the liquid phase pH value, arsenic species can be attached to one or two surface groups (monodentate and bidentate surface complexes similar to the mechanism on a weakly basic anion exchanger; Figure 3.2).

Activated alumina, $Al_2O_3/Al(OH)_3$ is one of the inorganic amphoteric exchange materials applied for As removal. It has a large internal surface in the range of 200–300 m^2 g^{-1}. Its pH_{PZC} is approximately 8.2. As a consequence it exhibits an increasing As capacity with decreasing pH value. Capacities of about 1.6 g L^{-1} of packed bed have been reported at pH 6 (Frank and Clifford 1990). As polymeric anion exchange resins, activated alumina exhibit a preference of ions according to a series of selectivity (Clifford 1999):

$$OH \gg HPO_4^{2-} > HAsO_4^{2-} > F^- > SO_4^{2-} \gg HCO_3^- > Cl^- > NO_3^-$$

As a consequence, arsenate species are highly preferred and major decreases of the sorption capacity are only due to phosphate while the competitive sorption of sulphate ions is less efficient. Regeneration can be carried out by NaOH solutions, followed by flushing with acid to re-establish the positive surface charge. Regeneration is more difficult and less effective compared to ion exchange resins and leads to the displacement of only 50–80% of As species (Clifford and Lin 1986; Driehaus 1999; Wang *et al.* 2000). Furthermore, there is some material loss due to the dissolution of activated alumina in high-alkaline media. Advantages of the application of activated alumina are the simple technology in which service cycles extend to the treatment of several thousand bed volumes of raw water before regeneration is required. Disadvantages include the relatively narrow pH range and the regeneration difficulties. Compared to polymeric ion exchangers, the sorption rate is considerably smaller and significantly longer contact times/smaller filtration rates are required (Anonymous 2000).

Much better efficiency has been found for iron oxide/hydroxide-based materials. Several different products have been developed, among which Granular Ferric Oxide and especially Granular Ferric Hydroxide, GFH®, outperformed any other sorbent materials. GFH® is a synthetic akaganeite material with a point of zero charge (PZC) at pH 7.6. This material has been tested in many laboratory, bench and pilot-scale experiments and is manufactured and applied on a large scale. Its advantage is that arsenate anions are strongly retained and, therefore, immobilised while small product water concentrations are achieved. Depending

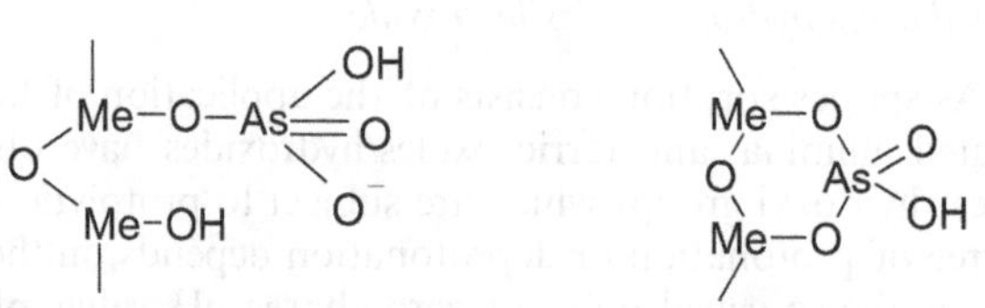

Figure 3.2.   Postulated monodentate (left) and bidentate (right) surface complexes at hydrous oxides. "Me" symbolises the metal (e.g., Al, Fe, and similarly Mn, Ti).

on the raw water concentration, AsO_4^{3-} capacities may amount to 28 g L^{-1}. At feed water concentrations of 10–50 μg L^{-1}, 50,000–300,000 bed volumes of raw water can be treated, depending on pH and phosphate concentration (Dlugosch 2001; Driehaus 2002; Pal 2001). Unlike activated alumina, GFH® is not regenerated but discharged to landfill deposits and replaced by fresh material. Because of the strong immobilisation, As species are not re-mobilised under natural conditions. One additional advantage, unlike with activated alumina, is that $As^{(III)}$ species are effectively removed – probably due to an oxidation of $As^{(III)}$ to $As^{(V)}$ at the surface. Granular Ferric Oxide and especially GFH® have been applied in numerous technical plants including household devices in West Bengal, India.

A third inorganic material that has successfully been studied for As-species removal is manganese dioxide, MnO_2 (Driehaus et al. 1995). Similar to iron oxide materials, As species are effectively eliminated and relatively large solid-phase concentrations can be achieved. Similar to GFH®, $As^{(III)}$ species are oxidised at the surface and then sorbed as $As^{(V)}$ species (Prasad 1994). Several other iron/manganese oxide-based materials have been investigated over the last 20 years, partly in the framework of field tests (Anonymous 2000; Deschamps et al. 2005; Prasad 1994; Shevade and Ford 2004; Zeng 2003). Among these materials are green sand and other natural minerals as well as iron-coated sand. The mechanisms of sorption are more or less identical to those described above. Some of them have to be regenerated.

One of the disadvantages of the granular hydrous metal oxides/hydroxides is the slow sorption rate. This can be overcome only by increasing the specific outer surface, using smaller particles. Because micro-particles do not allow a conventional filter operation, micro-particles with magnetic properties are being developed, together with the respective application technology (Dahlke et al. 2003). Another development is based on a polymeric network, comparable to that of ion exchange resins, in which nano-particles of iron oxide/hydroxide are introduced. These sorbents combine the efficient sorption onto ferric oxide/hydroxide material with the fast sorption onto nano-particles and the application of conventional packed beds (DeMarco et al. 2003; SOLMETEX 2004). Among the most recent investigations is the use of zerovalent Fe-particles for As sorption and removal (Manning et al. 2002).

3.3.3   *Macrophyte potential for As removal*

Phytoremediation, especially rhizofiltration has stimulated many researchers to investigate the potential of different aquatic plant species to remove potentially toxic elements such as boron (B), chromium (Cr), copper (Cu), mercury (Hg), cadmium (Cd), nickel (Ni), selenium (Se) and arsenic (As) in water bodies (Mkandawire and Dudel 2005). Metal sorption by plant cells is facilitated by mechanisms that involve transporter proteins. Due to the structural similarity between phosphate and arsenate, $As^{(V)}$ is sorbed by plants at the same site as phosphate. The translocation is done with protons H^+ (Mkandawire et al. 2004; ▶ 13).

Three aquatic macrophyte species (*Azolla caroliniana* Wild, *Savin minimum* Baker and *Lemna gibba* Linnaeus) have shown potential to remediate impacted aquatic environments (Guimarães 2006). The plants' removal efficiency from solutions decreases with increasing As availability. The three species efficiently removed arsenic from solutions when exposed to concentrations of 0.5 mg L^{-1}, with *L. gibba* showing the best removal potential.

3.3.4   *Heterogeneous photocatalysis with TiO₂ for As removal by co-precipitation with ferric sulphate*

So-called Advanced Oxidation Processes (AOP) have attracted great interest in environmental decontamination due to the formation of the hydroxyl radical (• OH), a highly oxidizing agent. Heterogeneous photocatalysis, a process involving redox reactions induced by radiation on the semiconductor mineral surfaces (catalysts), e.g., TiO_2, are one of these AOP. Some work related to the photocatalytic oxidation of arsenite to arsenate has been conducted using

TiO_2 in suspension. This is considered technically effective and environmentally acceptable to remediate As contaminated water.

Method development and implementation for As removal from surface and groundwater employ heterogeneous photocatalysis with TiO_2 that is immobilized inside a photochemical reactor to oxidize $As^{(III)}$ and subsequent removal of $As^{(V)}$ by coprecipitation with ferric sulphate. The higher the dosage of ferric ions, the more Fe-hydroxide precipitates are formed, leading to a greater surface area to sorb arsenate and consequently, to lower residual As concentrations (Mendes *et al.* 2009). The hydroxide yield depends on hydrolysis rate, solution pH, temperature, Fe-concentration and the presence of other anions. The efficiency of the coagulation process depends on the initial As concentration. The higher the initial $As^{(III)}$ concentration, the higher the amount of $As^{(V)}$. A higher residual Fe-amount is needed for complete As removal. Following the oxidation process in the reactor, the obtained $As^{(V)}$ concentration was about 9.8 times that of $As^{(III)}$. Tri-valent arsenic is more difficult to remove and is weakly sorbed to Fe-hydroxides. An oxidation step prior to $As^{(III)}$ to $As^{(V)}$ transformation to improve the efficiency of the sorption process and to decrease the Fe-amount used for the treatment of As contaminated water is more significant for the decay of $As^{(V)}$ as compared to $As^{(III)}$. The final steps to oxidate and remove arsenic (TiO_2 film prepared with a suspension of 10% w/v; pH: 7.0; oxidation time: 30 min; Fe^{3+}-concentration 50 mg L^{-1}) used water samples, supplemented with 1.0 mg $As^{(III)}$ L^{-1}, to verify the influence of the matrix. After treatment, more than 99% of the arsenic was removed from the water.

3.4   ARSENIC ELIMINATION THROUGH COAGULATION/FILTRATION

Coagulation/filtration is one of the classical possibilities to remove As species from water which has been studied in a large number of investigations. It has been applied to the treatment of drinking water in many cases. Coagulants can be alum $[Al_2(SO_4)_3]$, ferric chloride $[FeCl_3]$, or ferrous sulphate $[FeSO_4]$ (Baldauf 1995; Jekel and VanDyck-Jekel 1989; Seith and Jekel 1999). The mechanism of As elimination is that of a sorption onto the freshly precipitated $Al(OH)_3$ and $Fe(OH)_3$ particles or flocs. The results clearly indicate that the application of Fe-salts provides generally better elimination than Al-salts. Elimination was effective, when arsenic is present as $As^{(V)}$. Almost complete elimination has been observed at dosages >5 mg $FeCl_3$ L^{-1}. At raw water concentrations of 10–50 µg L^{-1}, a dosage of 1–2 mg Fe^{3+} L^{-1} leads to a satisfactory As elimination below the standard for drinking water (10 µg L^{-1}; ▶ 5). In contrast, elimination of $As^{(III)}$ species was rather unsatisfactory. However, oxidation by H_2O_2, NaOCl, or Cl_2 quickly converts $As^{(III)}$ to $As^{(V)}$ and also leads to an efficient $As^{(III)}$ elimination. Application of $FeCl_3$ generates relatively big flocs while finer flocs are formed from $FeSO_4$ (Baldauf 1995; Jekel and VanDyck-Jekel 1989; Seith and Jekel 1999). The disadvantage of the process lies in a relatively large volume of As-bearing sludge that needs to be discharged.

A similar reaction mechanism of As species with freshly precipitated hydrous oxides/hydroxides occurs during oxidation of iron and manganese from reduced groundwaters by aeration. Arsenic elimination is fairly efficient during Fe precipitation whereas it is less efficient during Mn-precipitation (Baldauf 1995). This kind of As elimination is common practice in many drinking water works using reduced groundwater as raw water. The same kind of As elimination has been proposed and studied in field experiments for an underground As elimination. In this case, reduced groundwater is pumped out, saturated with oxygen and re-infiltrated into the underground. Because of the oxygen, iron and manganese are oxidised and generate oxide and hydroxide flocs onto which $As^{(V)}$ species are sorbed and, thus, immobilised (Rott and Friedle 2000). Arsenic elimination can even be achieved by sorption onto freshly generated $Fe(OH)_3$ flocs from the corrosion of iron-based pipelines (Karschunke 2005).

3.5 ARSENIC ELIMINATION BY MEMBRANE PROCESSES

Membrane processes offer additional possibilities for As removal. Elimination can be achieved by i) filtration of As bearing particles, ii) exclusion because of size of hydrated ions, or iii) electric repulsion by the membrane (see references in Anonymous 2000).

Microfiltration and ultrafiltration do not allow any direct elimination because As species are far too small and can pass the membrane. With these processes, elimination is due to a physical sieving of As bearing particles. Microfiltration per se cannot remove any arsenic. However, it can be combined with coagulation by means of ferric chloride or alum and can, therefore, be applied (Anonymous 2000).

Ultrafiltration can remove colloidal constituents and therefore, can be applied if the raw water contains arsenic or As bearing species attached to such particles. Elimination can be enhanced using electrically charged ultrafiltration (UF) membranes. Comparison of some neutral and negatively charged membranes yielded unsatisfactory elimination rates of up to 63%. The results were confirmed during pilot scale experiments with the most promising UF membranes (Anonymous 2000).

Nanofiltration, which usually shows a predominant removal of divalent species, can eliminate $As^{(III)}$ and $As^{(V)}$ species predominantly through size exclusion. Therefore, it presents a reliable method for As elimination. Arsenic removal rates in bench and pilot scale experiments amounted to 60 to >95%. Again the results were less satisfactory for $As^{(III)}$, which may pass the membrane as a non-charged species. Because of the small pore size, however, nanofiltration membranes are subject to fouling at high DOC levels. This may cause reducing conditions at the membrane which leads to a change in speciation from $As^{(V)}$ to $As^{(III)}$ and to a substantial drop of the elimination performance (Anonymous 2000).

Reverse osmosis has been shown to be another reliable method for As elimination. Both bench and pilot-scale experiments demonstrated $As^{(V)}$-elimination rates >95% and $As^{(III)}$-elimination rates of about 74%. A slight increase of performance was observed for high DOC raw waters (Anonymous 2000).

Electrodialysis is a membrane process which removes ionic species by rejection at the membrane with the same charge sign as the ion. Electrodialysis with reversal of the polarity of the electrodes (EDR) has been investigated for As removal and has been studied in pilot-scale experiments. Because the separation effect is based on the presence of charged species, elimination of $As^{(III)}$ is generally poor. In pilot tests with raw water predominantly containing $As^{(III)}$, the As concentration was reduced from 188 to 136 µg L^{-1} (only 30%). The efficiency was better and amounted to 73% with a mixed $As^{(III)}$ and $As^{(V)}$ composition (Anonymous 2000).

REFERENCES

Anonymous: Technologies and costs for removal of arsenic from drinking water. International Consultants, Inc., Malcolm Pirnie, Inc., The Cadmus Group, Inc., 2000. EPA 815-R-00-028; Available at http://www.epa.gov/safewater/arsenic/pdfs/treatments_and_costs.pdf.
Baldauf, G.: Aufbereitung arsenhaltiger Wässer. *gwf Wasser Special* 136:14 (1995), pp.99–110.
Bothe, J.V. jr. & Brown, P.R.: Arsenic immobilisation by calcium arsenate formation. *Environ. Sci. Technol.* 33:21 (1999), pp.3806–3811.
Clifford, D.: Ion exchange and inorganic adsorption. In: Letterman, A. (ed): *Water quality and treatment*, Volume 9. AWWA, McGraw Hill, New York, 1999, pp.561–640.
Clifford, D. & Lin, C.C.: *Arsenic removal from groundwater in Hanford, California – a preliminary report.* University of Houston, Department of Civil / Environmental Engineering, 1986.
Dahlke, T., Holzinger, S., Chen, Y.H., Franzreb, M., Höll, W.H., Eldridge, R. & Nguyen, H. van: Development and application of magnetic micro sorbents for removal of undesirable ionic contaminants from waters. *Proc. CHEMCA 200* (31st Australasian Chem Engin Conf) Sept. 28 – Oct. 1, 2003, Adelaide, South Australia, 2003.

DeMarco, M.J., SenGupta, A.K. & Greenleaf, J.E.: Arsenic removal using a polymeric/inorganic hybrid sorbent. *Water Res.* 37:1 (2003), pp.164–176.

Deschamps, E., Ciminelli, V.S.T. & Höll, W.H.: Removal of As[(III)] and As[(V)] from water using a natural Fe and Mn enriched sample. *Water Res.* 39:20 (2005), pp.5212–5220.

Dlugosch, T.: *Development, testing and characterisation of manganese dioxide coated magnetic sorbents for the elimination of arsenic species from water.* Unpubl. Diploma thesis, IHEE, Delft, Forschungs-zentrum Karlsruhe, CSIRO Australia, 2001.

Driehaus, W.: Arsenentfernung aus Grundwasser und Trinkwasser mit dem GEH-Verfahren. In: Rosenberg, F. & Röhling, H.G. (eds): *Arsen in der Geosphäre, Schriftenreihe Deutsche Geol. Ges.* Volume 6, 1999, pp.133–137.

Driehaus, W.: Arsenic removal – experience with the GEH® process in Germany. *Water Sci. Technol., Water Supply* 2:2 (2002), pp.276–280.

Driehaus, W., Seith, R. & Jekel, M.: Oxidation of arsenate[(III)] with manganese oxides in water treatment. *Water Res.* 29:1 (1995), pp.297–305.

Frank, P. & Clifford, D.: *Arsenic[(III)] oxidation and removal from drinking water.* EPA/600/S-2-86/021; Water Engineering Research Laboratory, Cincinnati, 1990, p.9.

Guimarães, F.P.: *Potencial de macrófitas para remoção de arsênio e atrazine em solução aquosa.* Unpubl. M.Sc. thesis, Federal University of Viçosa, 2006, p.72.

Horst, J. & Höll, W.H.: Application of the surface complex formation model to ion exchange equilibria. IV: Amphoteric sorption onto γ-Aluminium oxide. *J. Coll. Interface Sci.* 195:1 (1997), pp.250–260.

Jekel, M. & Van Dyck-Jekel, H.: Spezifische Entfernung von anorganischen Spurenstoffen bei der Trinkwasseraufbereitung. *DVGW-Schriftenreihe Wasser* 62, 1989.

Karschunke, K.: Nutzung der Eisenkorrosion zur Entfernung von Arsen aus Trinkwasser. PhD dissertation, Technical University Berlin, 2005, p.171. http://deposit.ddb.de/cgi-bin/dokserv?idn=97502065x&dok_var=d1&dok_ext=pdf&filename=97502065x.pdf.

Lenoble, V., Laclautre, C., Serpaud, B., Deluchat, V. & Bollinger, J.C.: As[(V)] retention and As[(III)] simultaneous oxidation and removal on a MnO$_2$-loaded polystyrene resin. *Sci. Total Environ.* 326:1–3 (2004), pp.197–207.

Magalhães, M.C.F.: Arsenic. An environmental problem limited by solubility. *Pure Appl. Chem.* 74:10 (2002), pp.1843–1850.

Manning, B.A., Hunt, M.L., Amrhein, C. & Yarmoff, J.A.: Arsenic[(III)] and arsenic[(V)] reactions with zerovalent iron corrosion products. *Environ. Sci. Technol.* 36:24 (2002), pp.5455–5461.

Mendes, G., Bellato, C.R. & de Oliveira Marques Neto, J.: Heterogeneous photocatalysis with TiO$_2$ in the oxidation of arsenic and its removal from water by coprecipitation with ferric sulfate. *Química Nova*, 32:6 (2009), pp.1471–1476.

Mkandawire, M. & Dudel, E.G.: Accumulation of arsenic in *Lemna gibba* L. (duckweed) in tailing waters of two abandoned uranium mining sites in Saxony, Germany. *Sci. Total Environ.* 336:1–3 (2005), pp.81–89.

Mkandawire, M., Lyubun, Y.V., Kosterin, P.V. & Dudel, E.G.: Toxicity of arsenic species to *Lemna gibba* L. and influence of phosphate on arsenic bioavailability. *Environ. Toxicol.* 19:1 (2004), pp.26–35.

Nishimura, T., Itoh, C.T. & Tozawa, K.: Equilibria of the systems Co[(II)]-As[(III,V)]-H$_2$O and Ni[(II)]-As[(III,V)]-H$_2$O at 25°C. *Bull. Res. Inst. Min. Dress Metall, Tohoku Univ.* 49:1 (1993), pp.61–70.

Pal, B.N.: Granular Ferric Hydroxide for elimination of arsenic from drinking water. *Proc BUET-UNU Workshop "Technologies for removal of arsenic from drinking water"*, 5 May, 2001, Dhaka, Bangladesh, 2001, pp.59–68.

Prasad, G.: Removal of arsenic[(V)] from aqueous systems by adsorption onto some geological materials. In: Nriagu, J. (ed.): *Arsenic in the environment*, Volume I. John Wiley & Sons, New York, 1994, pp.133–154.

Riedel, E.: *Anorganische Chemie.* 3rd ed, W. de Gruyter, Berlin, New York, 1994.

Robins, R.G.: Some chemical aspects relating to arsenic remedial technologies. *Proc US EPA workshop on managing arsenic risks to the environment.* Denver, CO, (1–3 May 2001), 2001. http://www.epa.gov/ttbnrmrl/ArsenicPres/78.pdf.

Robins, R.G., Nishimura, T. & Singh, P.: Removal of arsenic from drinking water by precipitation, adsorption or cementation. *Proc BUET-UNU Workshop "Technologies for removal of arsenic from drinking water"* May 2001, Dhaka, Bangladesh, 2001, pp.31–42.

Rott, U. & Friedle, M.: 25 Jahre unterirdische Wasseraufbereitung in Deutschland – Rückblick und Perspektiven. *gwf Wasser-Abwasser* 141 (2005), pp.99–107.

Seith, R. & Jekel, M.: Aufbereitung arsenhaltiger Rohwässer zu Trinkwasser. In: Rosenberg, F., Röhling, H.G. (eds.): *Arsen in der Geosphäre, Schriftenr. Deutsche Geol. Ges.,* Volume 6, 1999, pp.55–66.

Shevade, S. & Ford, R.G.: Use of synthetic zeolites for arsenate removal from pollutant water. *Water Res.* 38:14–15 (2004), pp.3197–3204.

SOLMETEX: ArsenXNP Rio Rancho field trial report, 2004. Available at http://www.solmetex.com/ pdfs/RioRanchodata.pdf.

Tenny, R.: Ferric salts reduce arsenic in mine effluent by combining chemical and biological treatment. *Environ. Sci. Engin.* 14:1 (2001), pp.25–29.

Twidwell, L.G., McCloskey, J., Miranda, P. & Gale, M.: Technologies and potential technologies for removing arsenic from process and mine wastewater. *Proc REWAS99,* 5–9 Sept. 1999, San Sebastian, Spain, 1999.

Vogels, C.M. & Johnson, M.D.: Arsenic remediation in drinking waters using ferrate and ferrous ions, Technical Completion Report, Account No. 01-4-23922, New Mexico Resources Research Institute, 1998.

Wang, L., Chen, A. & Fields, K.: Arsenic removal from drinking water by ion exchange and activated alumina plants, 2000. EPA/600/R-00/088.

West General, 2006. Available at http://www.westgeneral.com/outofthebox/compounds/assol.html.

Zeng, L.: A method for preparing silica-containing iron$^{(III)}$ oxide adsorbents for arsenic removal. *Water Res.* 37:18 (2003), pp.4351–4358.

CHAPTER 4

Technology options for arsenic removal and immobilization

Eleonora Deschamps

4.1 INTRODUCTION

Arsenic contamination of ground and surface waters is of a similar order of magnitude in Latin America as in other world regions (▶ 1). However, the selection of the best technology for As removal from drinking water poses a challenge for researchers due to the large number of factors to be considered.

An extensive list of scientific publications exists on As-removal from water. In theory this removal can be achieved through different established treatment methods. Some classical treatment processes are applied on the industrial scale (▶ 3). It is noteworthy that only a few of those technologies have been adapted to smaller scales (local systems). The lack of funding and technical assistance has slowed down commercial development of low-cost technologies for small communities (Litter *et al.* 2010). No doubt, the control of interfering variables and factors make the application of the methodology on small treatment systems much more difficult, as these infrastructures are usually simple and human resources not properly qualified. Technology selection should be based on several factors such as:

• Initial and final As-concentration
• Amount of water to be treated
• Knowledge of the As-species present
• Available analytical methods
• Treatment plant's operational and maintenance costs
• Operator's degree of skills
• Management of the As-containing sludges, liquid or solid residues
• Quality of the raw water matrix. This requires a detailed knowledge of its physico-chemical composition and of economical constraints, among others.

In some Latin American areas, including Brazil, ground and surface waters are affected by mining activities and may contain high concentrations of Fe, Mn, and trace elements. Such contamination, just like a high mineralization or any other raw water anomaly, directly affects the performance of a water treatment system – and therefore needs to be understood in more detail. While As levels may be the same in different countries or regions, other compounds may be present in highly variable concentrations (Sancha 2003). As an example, $As^{(III)}$ represents 70–90% of total arsenic found in Tawain, where water contains high levels of natural organic matter, whereas water in Chile is characterized by the absence of organic matter, by high salinity and hardness and by high silica levels and oxidizing conditions, favouring the presence of $As^{(V)}$.

Technologies need to be selected that reduce arsenic and other toxic components to the safe levels established by legislation (▶ 5). While this should be self-explanatory, it is often not, and is of particular relevance in all countries with increasing drinking water shortages. Due to serious As-related risks for human health (▶ 2), the search has increased for processes capable of removing $As^{(V)}$ and $As^{(III)}$ and of keeping the element immobilized, and thus limiting its bioavailability (Deschamps *et al.* 2003). Arsenic mobilization from solids into water is mainly controlled by local redox, geomorphological, geological, hydrological and biogeochemical conditions. On the other hand As immobilization usually involves geochemical

49

conditions (pH and Eh), adsorption to different substrates or co-precipitation with Fe or Mn-oxihydroxides.

In Brazil, arsenic is predominantly released from both past and present gold mining activities and occurs when liberated by weathering in hydrosphere and biosphere (Bundschuh *et al.* 2008). Groundwater and surface water resources are mainly affected by weathering of sulphidic ore deposits accompanied by As, Fe, Mn, Zn, Pb, Cd, Ni and other potential contaminants found at mining sites. Despite a large number of data and many studies by local researchers, these data often remain unpublished in refereed literature. This is also true for river pollution and jeopardized natural water quality. All monitoring points in the das Velhas river basin in Minas Gerais show As values above the drinking water limit of 10 mg L^{-1} (IGAM 2010; ▶ 11).

4.2   ARSENIC REMOVAL TECHNOLOGIES

Available technologies for As removal include traditional technologies ranging from adsorption and/or co-precipitation with metal hydrolysates, such as aluminium and iron; adsorption with natural and synthetic materials such as activated alumina, Fe and Mn-oxihydroxides; use of ionic exchange resins and membrane processes, such as reverse osmosis; electrodialysis and nanotechnologies, to emerging technologies, e.g., based on biological processes, solar and desalinization technologies (▶ 3). In general, these technologies successfully remove arsenic, especially when present in the pentavalent form. However, when a low drinking water standard is set, effective As removal from the water requires a complete oxidation of As$^{(III)}$ to As$^{(V)}$ (Khoe *et al.* 2000).

Removal by coagulation and co-precipitation with iron and aluminium salts. The most common technology for As removal from drinking water is conventional coagulation and co-precipitation. It is widely used for treating large water volumes. Treatment consists of coagulation, followed by flocculation, decantation and filtration. The coagulation may occur with the addition of either aluminium salts or iron salts (WHO 2002 points to greater efficiency with Fe salts). As the solubility of Fe-compounds is higher than that of aluminium, all iron in the water is converted into Fe-hydroxide, an As-sorbent. This does not occur with aluminium, however, a fact which may explain its lower efficiency in As removal.

Usually, the efficiency of this As-removal route is a function of (a) its oxidation status, (b) the quantity and form of the iron present, (c) the type and doses of the coagulant used, (d) water pH, (e) the initial As concentration, (f) the existing competitor ions (such as phosphate, silicate and organic matter), (g) the agitation speed, and (h) the filter backwash frequency (EPA 2004).

It is important to note that As$^{(V)}$ is more efficiently removed than As$^{(III)}$, which usually requires an oxidative pre-treatment to improve efficiency. As the oxidation by O$_2$ and sunlight has a very low reaction rate, oxidation must be enhanced by the addition of free chlorine, ozone, or potassium permanganate. Sancha *et al.* (1992) suggested oxidation through the free chlorine as an initial step, which has the additional advantage of avoiding biological growth in the filter medium (disinfection).

This conventional route for As removal is highly efficient when applied to the treatment of surface water with high turbidity, from which not only arsenic is removed but other contaminants as well, resulting in the most efficient As removal in the decantation step. The main disadvantage of this procedure is the generation of relatively large volumes of As-containing sludge, which must be disposed as a hazardous residue.

For better quality and low turbidity ground waters, this conventional treatment is not recommended due to the complexity of the operation in this situation, the amount of coagulant, the volume of sludge produced and the related plant costs. In this case, As removal by ion exchange resin or even by sorption with activated alumina is usually more efficient, albeit more expensive. If the reagents (Fe and Al salts) can be provided at low cost, this option, scaled-down for domestic applications, can be attractive due to its simple and basically maintenance-free operation (Litter *et al.* 2010).

Removal by ion exchange resin. Ion exchange is a fixed bed process that involves the exchange of dissolved ions inside a resin. The efficiency of As removal by ion exchange is strongly affected by the As-ionic species and the water matrix quality. Non-charged As-species H_3AsO_3 predominate in the pH-band of natural waters, and since the ion exchange resin preferably removes anionic species, it is necessary to first oxidize As$^{(III)}$ to As$^{(V)}$ to achieve the desired removal efficiencies.

Competitor ions at the exchange sites make the ion exchange resins economically unattractive for waters containing total dissolved solids >500 mg L^{-1} or sulphates >500 mg L^{-1}. After exhaustion, the resin should be regenerated with a solution of sodium chloride. The resulting brine classifies as hazardous waste, which is a drawback of this process. It is also necessary to properly dispose of the resin once its adsorptive capacity is exhausted (Pires *et al.* 2005; Wang *et al.* 2000).

Removal by activated alumina and other aluminium materials. Granular activated alumina was the first successfully applied sorbent for As removal in water treatment systems. As with the use of ion exchange resin, the efficiency of the As removal by activated alumina is strongly affected by the aqueous As-species and by the water matrix quality. As$^{(V)}$ adsorption of activated alumina is highly dependent on water pH, which requires adjustment, but is little affected by competitor ions. If the As$^{(III)}$ removal efficiency is low, it is necessary to include a pre-oxidation step in the treatment. Once its adsorptive capacity is exhausted, a significant disadvantage is that it must be disposed of as a hazardous residue (Wang *et al.* 2000).

Esper *et al.* (2007) have studied and compared the use of Al and Fe-oxisols in Brazil as a liner in a tailings dam to retain arsenic. Goethite was superior to gibbsite in respect to As immobilization. This study showed that a combined enriched Fe and Al-oxisol liner is an efficient system on an industrial scale. On a small and household-scale application, Lujan and Graieb (1995) showed the suitability of an activated aluminium hydroxide hydrogel to remove arsenic, however, it is important to mention that costs can be a limiting factor.

Removal by reverse osmosis, electrodialysis and nanotechnology. Arsenic removal by reverse osmosis and nanotechnology strongly depends on chemical water characteristics. The presence of calcium ions and organic matter in suspension presents a limiting factor that needs to be addressed. As a result of this, water needs to be preconditioned before treatment. In ground waters, the presence of humic substances and other organic and inorganic compounds such as methane, sulphides, and ammonium are also important removal-limiting factors. It should be noted that reverse osmosis consumes between 20 and 50% of the raw water. This adds to the high energy consumption and may render the application of this technology unfeasible in arid regions (Sancha and Fuentealba 2009). However, the advances in membrane technology have reduced the costs of reverse osmosis (lower pressure and lower energy requirements) especially when combined with available renewable energy resources, and have turned into a viable choice for the production of potable water (Bundschuh *et al.* 2010).

In pilot tests with electrodialysis, where As-removal is based on electrical charge, it has been observed that As$^{(III)}$ removal is rather limited even in the presence of As$^{(V)}$ (Cammarota *et al.* 2001). The removal efficiency is around 80% and the recovery of the treated water is only 20–25%, which makes electrodialysis unfeasible in regions with water shortage. Comparing this technology with the reverse osmosis and nanotechnology, electrodialysis is not competitive due to the high cost and low efficiency (EPA 2004).

In situ and biological treatment removal. Arsenic in groundwater and therefore often under reducing conditions, occurs in its most mobile species, As$^{(III)}$. Under aerobic condition, arsenic may be immobilized by sorption and/or co-precipitation to Fe-oxides. This technique is known as in situ removal (Johnston *et al.* 2001). Experiments in groundwater in Germany, rich in iron and manganese and using atmospheric oxygen injections, achieved As levels up to 5 µg L^{-1} (Rott and Friedle 2001). Another process consists in the oxidation of Fe$^{(II)}$ in the groundwater via bacteria, with subsequent As-sorption or co-precipitation. The in situ and biological treatment removal techniques are in an experimental phase and need more thorough studies.

Removal by adsorption. Arsenic removal by sorption has been highlighted as well as its increasing application in water treatment plants. Sorption occurs in a fixed bed process through which ions in solution are removed by sorption at sorbent active sites. When all active sites are occupied, the sorbent should be regenerated or replaced and properly disposed. Granular activated alumina was the first sorbent successfully applied in As removal treatment systems. Other sorbents, mostly Fe-based, were developed and made available on the market.

Differences in the general composition of sorbents, such as: Fe valence state, crystalline structure, specific surface area, and other physico-chemical characteristics, result in different kinetics and sorption capacities. Studies and applications performed in commercial experiments have shown that the removal capacity of Fe-based sorbents exceeds that of alumina-based ones (EPA 2003). The performance of the treatment systems depends on various factors including concentration and speciation, pH, presence of competitor ions, adsorbent specific characteristics, average life and the extent to which they can be regenerated:

a. *Arsenic concentration and speciation*. The total As concentration in water affects the sorbent replacement costs. The sorbent removal capacity is increasingly exhausted with the increase of initial As concentrations. The proportion of $As^{(III)}$ and $As^{(V)}$ in water determines the need for oxidizing pre-treatment to assure system efficiency. $As^{(V)}$ is usually adsorbed more efficiently. Chlorine, ozone and potassium permanganate are the most effective oxidants for $As^{(III)}$ and $Fe^{(II)}$.

b. *pH-value*. Arsenic removal capacity can be improved with pH adjustment for both the alumina based and the Fe-based sorbent. Alumina-based sorbents remove mainly $As^{(V)}$ and the optimum pH-value is 5.5. The Fe-based sorbents have an affinity to both $As^{(V)}$ and $As^{(III)}$. However, $As^{(III)}$ adsorption is affected differently from $As^{(V)}$, depending on pH-value. $As^{(V)}$ adsorption decreases with increasing pH. The sorbent surface becomes less positive with pH and thus decreases the attraction for negatively charged As species ($H_2AsO_4^-$ and $HASO_4^{2-}$). $H_2AsO_4^-$ dominates for pH-values between 7 and 11. On the other hand, $As^{(III)}$ adsorption increases with pH-increase, with a maximum at pH 9. In natural water pH-ranges, non-charged As species (H_3AsO_3) predominate and $H_2AsO_3^-$ predominates at pH 9, increasing the affinity by the positively charged solid surface. The optimal pH-range varies from 5.5–9.0 for different Fe-based sorbents.

c. *Competitor ions*. The As adsorption is influenced by the presence of competitor ions for the active sites in the water, such as fluorides, bicarbonates, sulphates, silica and phosphates. Anions compete directly for the active sites and can change the electrostatic charge of the sorbent surface and affect the As removal effectiveness. For Fe-based sorbents, As removal capacity is reduced for silica concentrations >40 mg L^{-1}, phosphate concentrations >1.0 mg L^{-1} and sulphate concentrations >150 mg L^{-1}.

d. *Regeneration*. Some sorbents may be regenerated through chemical pH-adjustment after the exhaustion of As adsorption capacity. However, chemical regeneration usually is economically feasible only for large treatment systems and high As concentrations. For smaller systems the sorbent disposition should be the most appropriate option.

Saha *et al.* (2001) have studied the As removal by 18 different sorbents, among these sand, activated carbon, bauxite, hematite, Fe-oxide covered sand, activated alumina and hydrated Fe-oxide. They noted that most sorbents had low removal efficiency, and only three – activated alumina, Fe-covered sand and granulated hydrated Fe oxide – were effective for $As^{(V)}$ removal (over 90% for an initial As concentration of 1 mg L^{-1} in water and about 86% for $As^{(III)}$ at neutral pH – indicating the need for the $As^{(III)}$ pre-oxidation). A bibliographical review shows that $As^{(V)}$ adsorption is favoured at pH ranges between 4 and 7, when the sorbent surface has a positive charge and arsenic occurs in the anionic form. Adsorption occurs as a result of the active sites as well as due to electrostatic interaction (Edwards 1994). The superposition of the pH-Eh diagrams of the Fe species and the As species highlight the $As^{(V)}$ affinity zone with Fe salts, corresponding to positive Eh-conditions and a pH between 3 and 8 (Deschamps *et al.* 2003). The presence of phosphate (>10 mg L^{-1}) and fluorides (>2 mg L^{-1}) decreases removal effectiveness, while the presence of nitrate, sulphate, chloride, chromate, calcium, magnesium and iron does not significantly affect removal (Saha *et al.* 2001). The best results were achieved with granulated hydrated Fe-oxide and Mn-containing adsorbents. The results obtained with the Fe-covered sand were positive, since this adsorbent does not require pH control. However, its main disadvantage is that the sand activation with Fe-oxide process is rather complex.

Deschamps *et al.* (2003) studied As adsorption in natural samples of iron and Mn-oxides and hydroxides, and observed the advantages of having minerals able to oxidize and adsorb arsenic in the same sample. The tests demonstrated the high adsorption capacity on both the laboratory batch and the continuous scale, but was lower than that obtained with synthetic granulated hydrated Fe-oxide.

4.3   OVERRIDING FACTORS IN TECHNOLOGY SELECTION

Water quality. Mostly due to the high As affinity with iron, the Fe level in the water matrix is a decisive factor in the selection of an optimum treatment technology. Figure 4.1 illustrates the strong dependence of the technology on initial As and Fe concentrations in the water matrix, divided into three categories (A–C).

A. *High Fe levels.* Waters containing relatively high Fe levels (Fe/As ratio 20:1 or higher, EPA 2004) are those that best fit the Fe-based removal treatment (field A in Fig. 4.1). Hence, up to 50 µg As L^{-1} can be removed from waters with 1 mg Fe L^{-1}. This is a guide value and can be achieved under optimum adsorptive and operational conditions only.
B. *Moderate Fe levels.* For Fe/As ratios below 20:1, adsorption and/or co-precipitation with addition of Fe salts should be chosen (field B in Fig. 4.1).
C. *Low Fe levels.* For Fe levels below 300 µg L^{-1}, processes such as adsorption and membrane processes are the most appropriate (field C in Fig. 4.1). Problems may occur above 300 µg L^{-1}, such as water odour and colour and the likelihood of stains by Fe particulates in system components. The ionic exchange resins do not remove $Fe^{(III)}$ and As complexes efficiently (EPA 2004).

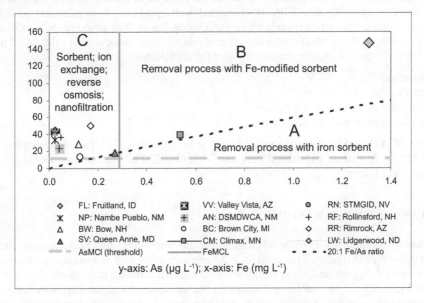

Figure 4.1.   Technology dependence of As and Fe concentrations in water matrix (EPA 2004).

Residue generation and disposition. All As-removal technologies produce residues that need to be disposed of, whether they are solid as the sorbent itself or liquid as the backwash water, and/or the brine from the ion exchange regeneration process. These residues have to undergo the Toxicity Characteristic Leaching Procedure (TCLP), defining how a residue is to be disposed of. The generation of liquid residues depends on flow and backwash frequency. Usually, the systems are backwashed once or twice a month in the adsorptive processes, with a production of 10–15 effluent bed volumes for each backwash. The backwash should be performed at even higher frequencies (daily or at least weekly) in the adsorption and/or co-precipitation process. Turbidity, As, Fe, Mn and total dissolved solids of the generated effluent should be controlled to define the disposal mode. The availability of a treatment plant or other disposal facilities should be taken into account, and have to comply with respective regulations.

Operation system complexity. Complex systems require more experienced operators, a particular challenge for small operation systems. When choosing a treatment technology, the operators often discuss backwash frequency, the need for chemical products (pH adjustment, chlorine addition, etc.) and the mean replacement frequency. The automation level available for system operation and data collection can reduce complexity and save time. The adsorption process offers several advantages for small systems. In general, operation is simplified, maintenance and cost requirements are low, and residues have to be properly disposed of.

Costs. With the usual resource limitations, operational costs are always a bigger issue for most small treatment systems than for large ones, and therefore, a decisive factor for technology selection. A strong correlation exists between the total costs (purchase and running) and the size of the treatment system. Systems with Fe-based adsorption processes yield the lowest costs (EPA 2004).

Other factors. The adaptability of the selected process to future expansions and to adaptations to technology improvements is another argument that influences process selection. It is important to consider whether the system can be upgraded in the future, e.g., with new and less costly options to replace the adsorbent.

4.4   EPA COMPARATIVE STUDY TO CHOOSE THE BEST TECHNOLOGY

After evaluating 70 technological proposals for As removal in drinking water, the U.S. Environmental Protection Agency (EPA 2004) selected twelve proposals to implement twelve water treatment units at the production scale in the United States. The objective was to evaluate treatment systems for small communities, in compliance with the new limit value of 10 µg L^{-1} (EPA 2004). Four treatment technologies were finally selected:

- Adsorption
- Ion exchange
- Adsorption and/or co-precipitation
- Modified system with Fe addition.

Levels of As$^{(III)}$ and As$^{(V)}$ in the water varied from 14–146 µg L^{-1}, Fe$^{(II)}$-levels varied from <25–1.325 µg L^{-1}, while pH stayed between 7.2 and 8.5. The system flow varied between 186 and 2.912 L min^{-1}.

4.5   FINAL CONSIDERATIONS

The most important factors to be taken into account when selecting the As removal method are: 1) water matrix quality, 2) desired final As-concentration and 3) system support capability. The availability of analytical competence to control removal process efficiency and of qualified labour force to control and maintain the implemented technology also needs to be considered. The current literature offers a wide range of alternatives, but few can actually be adopted at a field scale. It is important to seek the balance between the technological and the economical factors. Each of these technologies has advantages and disadvantages, which need to be investigated and balanced when selecting the appropriate drinking water treatment technology.

For most small systems, the adsorption and/or co-precipitation technology with Fe and Mn-based materials (modified to small scale) are the technologies of choice on a community level. The Fe-based adsorption systems have the lowest costs, are more effective for small size units, and are usually operated easily, but require careful disposal of the solid residues. The activated alumina adsorption process requires chemical handling, which makes this process too complex for most small systems. Besides, the activated alumina process may be non-effective in the end, as its adsorptive capacity decreases drastically with each regeneration cycle associated to the generation of the effluent to be disposed of. For the ion exchange process, the disposal of effluents with high As concentrations is a problem. The sulphate concentration in the water matrix affects each cycle extension. This process is, therefore, recommended for systems with low sulphate concentrations and total suspended solids and with a polishing step after filtration. The reverse osmosis process can be applied in combination with local renewable energy resources.

In Minas Gerais, Brazil, a small community with about 200 inhabitants used alternative water resources until 2006. It was untreated local water, transported by simple tubes, to supply the population with sufficient water (quantity wise; but with questionable quality; ▶ 16). To mitigate the As problem and the presence of Escheria coli in the water, a small treatment plant was build, based on the adsorption technique, using a synthetic Fe sorbent followed by a local low-cost sorbent that can be handled and maintained by the local population.

A review of the state of the art indicates that the selection of a technology for As removal in developing countries demands a special approach for each location and water type. Acceptance of an affected local community, site characteristics, and available finances need to be considered prior to implementation. A cost/benefit analysis is needed in pursuit of a solution to any public health problem. Experience and expertise is available, particularly in Latin America countries, but limited financial resources and public policies still hamper implementation and maintenance.

The most affected communities are generally those in rural areas, where water is not treated and whose populations are unaware of the related exposure risks, be it to arsenic or any other organic or inorganic potential toxin. It is crucial in these communities that state authorities (health, environment, planning, water supply services etc.) promote programs to prevent and control the human health risk posed by the consumption of water. Such programs should involve not only the local health and environment authorities but the affected communities as well – a participatory approach (▶ 9, 15).

REFERENCES

Bundschuh, J., Pérez Carrera, A. & Litter, M. (eds): *Distribución del arsénico en las regiones Ibérica e Iberoamericana.* CYTED, Argentina, 2008, p.230. ISBN 13978-84-96023-61-1. http://www.cnea.gov. ar/xxi/ambiental/iberarsenic.

Bundschuh, J., Litter, M., Ciminelli, V.S., Morgada, M.E., Cornejo, L., Hoyos, S.G., Hoinkis, J., Alarcón-Herrera, M.T., Armienta, M.A. & Bhattacharya, P.: Emerging mitigation needs and sustainable options for solving the problems for rural and isolated urban areas in Latin America – a critical analysis. *Water Res.* 44:19 (2010), pp.5828–5845. Available at http://www.ncbi.nlm.nih.gov/pubmed/20638705.

Cammarota, F.C.L., Schreier, J., Cipriani, M.J.I. & Ferreira, F.S.S.: Tratamento de águas residuárias provenientes de indústria de vidro por processos de membrana visando o reuso. In: 21. *Congresso Brasileiro de Engenharia Sanitária e Ambiental* 1, 1–10, 2001.

Deschamps, E., Ciminelli, V.S.T., Weidler, P.G. & Ramos, A.Y.: Arsenic sorption onto soils enriched with manganese and iron minerals. *Clays Clay Minerals* 5:2 (2003), pp.197–204.

Edwards, M.: Chemistry of arsenic removal during coagulation and Fe-Mn oxidation. *J. Am. Water Works Assoc.* 86:9 (1994), pp.64–78.

EPA: Minor clarification of the national primary drinking water regulation for arsenic. Fed Register 40 CFR Part 141. US Environmental Protection Agency. 2003. Available at http://www.epa.gov/fedrgstr/EPA-WATER/2003/March/day-25/w7048.htm.

EPA: Capital costs of arsenic removal technologies: US EPA Arsenic removal technology demonstration program Round 1, R-04/201: p.600. United States Environmental Protection Agency. 2004. Available at http://www.epa.gov/nrmrl/pubs/600r04201/600r04201.pdf.

Esper, J.A.M.M., Amaral, R.D. & Ciminelli, V.S.T.: Cover design performance at a Kinross gold mine in Brazil. In: *XXII Encontro Nacional de Tratamento de Minerios e Metalurgia Extrativa, Proceedings VII Meeting of the Southern Hemisphere on Mineral Technology.* Ouro Preto, Volume II. pp.607–612.

IGAM: Monitoramento da qualidade das águas superficias do Estado de Minas Gerais. Instituto Mineiro de Gestão das Águas, Belo Horizonte, 1, p.93.

Johnston, R., Heinjnen, H. & Wurzel, P.: Arsenic in drinking water. *Safe Water Technol.* 6 (2001), p.31., WHO, Final Draft.

Khoe, G.H., Emett, M.T., Zaw, M. & Prasad, P.: Removal of arsenic using advanced oxidation processes. In: Young, C.A. (ed): *Minor metals SME*, Littleton, CO, USA, 2000, pp.31–38.

Litter, M.I., Morgada, M.E. & Bundschuh, J.: Possible treatments for arsenic removal in Latin American waters for human consumption. *Environ. Pollut.* 158:5 (2010), pp.1105–1118.

Lujan, J.C. & Graieb, O.J.: Elimination of arsenic from water by distillation on a household scale in rural areas (in Spanish). *Rev Ciencia y Technológia* 3:7 (1994), p.13, Universidad Tecnologia Nacional, Tucumán, Argentina.

Pires, J.A., Dutra, A.J.B., Peres, A.E.C. & Martins, A.H.: Remoção de arsênio e metais pesados em solução aquosa empregando pisolito como sorvente natural. In: *XXI Encontro nacional de tratamento de minérios e metalurgia extrativa*, 2 (2005), pp.205–211.

Rott, U. & Friedle, M. Eco-friendly and cost-efficient removal of arsenic, iron and manganese by means of subterranean ground-water treatment. *Water Supply* 18:1/2 (2001), pp.632–636.

Saha, J.C., Dikshit, A.K. & Bandyopadhyay, M.: Comparative studies for selection of technologies for arsenic removal from drinking water. In: Ahmed, M.F., Ashraf, A.M., Adeel, Z. (eds): *Technologies for arsenic removal from drinking water BUET-UNU International workshop on technologies for arsenic removal from drinking water*, United Nations University, Tokyo, Japan, (2001), pp.76–84.

Sancha, A.M.: Removing arsenic from drinking water: a brief review of some lessons learned and gaps arisen in Chilean water utilities. In: Chappell, W.R., Abernathy, C.O., Calderon, R.L. & Thomas, D.J. (eds): *Arsenic exposure and health effects*, Volume V. Elsevier, Amsterdam, 2003, pp.471–481.

Sancha, A.M. & Fuentealba, C.: Application of filtration processes to remove arsenic from low-turbidity waters. In: Bundschuh, J., Armienta, M.A., Bhattacharya, P., Matschullat, J. & Mukherjee, A.B. (eds): *Geogenic arsenic in groundwater of Latin America*. In: Bundschuh, J. & Bhattacharya, P. (series eds): *Arsenic in the environment*, Volume 1. CRC Press/Balkema Publisher, Leiden, The Netherlands, 2009, pp.687–697.

Sancha, A.M., Rodrigues, D., Veja, F., Fuentes, S. & Lecaros, L.: Arsenic removal by direct filtration. An example of appropriate technology. In: *Internat. Seminar Proc. Arsenic in the environment and its incidence on health.* Univ de Chile, Santiago, Chile, 1992, pp.165–172.

Wang, L., Chen, A., Sorg, T. & Fields, K.: Field evaluation of arsenic removal by IX and AA. *J. Am. Water Works Assoc.* 4 (2000), p.94.

WHO: Guidelines for drinking-water quality. Arsenic in drinking water. Fact sheet N°210, World Health Organization, Geneva, Switzerland, 2002.

CHAPTER 5

Environmental law and regulations

Raquel Vieira, Sandra Oberdá & Nílton Rocha

5.1 INTRODUCTION

Worldwide concern about negative consequences of intolerable As levels for human beings or the environment triggered the establishment of international quality standards. Instead of following a precautionary principle, the action was directly linked to various types of environmental impact, as mankind reacts mostly on specific triggers only. Apart from negative effects studied in isolation and related to human health (▶ 2, 14), degraded areas generally show negative environmental effects. Both experiences lead to the current implementation of international public policies aimed at normalizing the exact threshold for tolerance or resilience towards arsenic, regardless of whether this is detrimental to man or to the environment, or the experienced changes are due to activity performed by man or by natural sources. In this context, the contact of river and spring waters with rocks that yield high As concentrations is an example of metalloid release from natural sources, while As pollution from mining and industrial activities stand for potentially detrimental alteration by anthropogenic activities (▶ 1, 11). Regardless of the circumstances, the introduction of limits provides a definition of safe health levels for everyone.

The first known As limit for drinking water was established in 1943. The United States of America set a value of 50 µg L^{-1} through their Environmental Protection Agency (EPA). Chile defined a value of 120 µg L^{-1} for drinking water through the Instituto Nacional de Normalización (INN) in 1970. The EPA ratified the value of 50 µg L^{-1} in 1975. Studies of that time, published in the international literature, indicated that this was a safe limit (Valencia 1999). To establish a safe limit, EPA in 1983 recommended considering the As valence states that strongly influence As toxicity and its health effects (▶ 2). The agency recommended including nutritional needs, because some scientific publications showed that arsenic is an essential nutrient for some vegetables and animal species, and that this metalloid might also be an essential nutrient for humans (Valencia 1999). In 1983, Chile assumed the limit of 50 µg L^{-1}, established by EPA. In 1985, EPA again confirmed the enforced limit of 50 µg L^{-1} (Valencia 1999).

The Bangladesh event was a regulatory turning point for the World Health Organization (WHO). An entire population was threatened by As intoxication due to the ingestion of contaminated water. The ongoing incident was likely triggered by a decrease of the water table, which led to weathering of As-containing pyrites. WHO defined a safety standard of 50 µg As L^{-1} for drinking water in 1984. Based on new toxicological evidence, WHO modified this value in 1993 to 10 µg L^{-1}. It is important to note that rules and standards, set by international organizations, may or may not be adopted by countries, leaving to non-signatory countries the option to adopt more permissive or restrictive limits. In this context, Germany, Japan, USA, Finland, Mongolia and Jordan adopted the WHO standard of 10 µg L^{-1}, each at their own time. Adopting a more permissive limit, Canada follows a standard of 25 µg L^{-1}, while India, West Bengal, Bangladesh, China, Vietnam and Chile kept the right to adopt the limit of 50 µg L^{-1}.

5.2   THE BRAZILIAN SITUATION

In Brazil, the incorporation of decisions by international organizations, such as WHO or UNEP, undergoes procedures similar to those related to treaties. The Brazilian State will express its consent in two distinct phases, 1) by adhesion to the standard, deliberated by the international organization, and 2) by issuing the referendum from the National Congress, according to the clause VIII, article 84 of the 1988 constitution. It should be noted that each Brazilian state adopts particular procedures related to the incorporation of rules from treaties or conventions. All standards relate to water potability, regardless of origin (ground or surface water). Until the year 2000, the Brazilian Ministry of Health defined a permissible maximum As concentration of 50 μg L^{-1}. Starting December 29, 2002, article 14 of the administrative rule GM n. 1.469, of 29.12.2000, established the value of 10 μg L^{-1}, accepting WHO's standards. This administrative rule sets more stringent potability standards and includes a larger number of parameters. It also determines the procedures and responsibilities related to the quality control of water for human consumption.

On March 26, 2004, the Ministry of Health published the administrative rule GM n. 518, of March 25, 2004, in the Diário Oficial da União that revoked GM n. 1.469/00. However, the potability standards to be met for drinking water for human use were left unchanged; thus the maximum value of 10 μg As L^{-1} has been maintained. In addition to the potability standards, the national environment policy permits the establishment of criteria and standards for environment quality and rules related to the use and handling of natural resources, which imply different benefits to human health (federal law n. 6.938 of 31.08.1981). The instruments in article 9 of this law are as follows:

  I establishment of environment quality standards,
  II environmental zone demarcation,
  III evaluation of environmental impacts,
  IV licensing and review of effective or potentially pollutant activities,
  V incentives to the production and implementation of equipment and the creation or absorption of technology to improve environmental quality,
  VI creation of protected spaces, relevant for ecological interests, environmental protection, and extractive reserves,
  VII establishment of a national environmental information system,
  VIII establishment of a federal technical register of activities and instruments for environmental defense,
  IX disciplinary penalties or compensation to comply with the measures necessary to preserve and restore environmental degradation,
  X annual publication of the Quality Report on the Environment, by the Brazilian Institute of Environment and Natural Resources (IBAMA),
  XI establishment of a governmental guarantee to provide environmental information,
  XII establishment of a federal technical register of potentially polluting activities and/or users of environmental resources.

Clause I of article 9 transcribed herein refers to the definition of environmental quality standards that express the quality level or grade, elements, relationships or sets of components, usually established in numbers, which meet particular functions, purposes or objectives, and are accepted by society. With this purpose, the Environment National Council (CONAMA) approved and published resolution n. 20 on July 18, 1986. It sets 50 μg L^{-1} as the limit and conditions for As release, irrespective of water stream classes. Recently, the above mentioned resolution was totally revoked by the CONAMA resolution n. 357 (17.03.2005), which edits the standards (Table 5.1).

Table 5.1.   Standards for maximum permissible As release in water ways.

Classification	Limit (μg L^{-1})
Class 1. Freshwater	10
Class 1. Freshwater with fishing or aquaculture for intensive consumption	140
Class 3. Freshwater	33
Class 1. Saline water	10
Class 1. Saline water with fishing or aquaculture for intensive consumption	140
Class 2. Saline water	69
Class 1. Brackish water	10
Class 1. Brackish water with fishing or aquaculture for intensive consumption	140
Class 2. Brackish water	69

CONAMA (2005).

Effluents from any polluting source may be released directly or indirectly into water courses, as long as they comply with the standard of 50 μg L^{-1}. The Council has not defined the As standards in groundwater, or in soil, vegetables, sediments and dust, which leads to diversified treatment within the federal state. The Technology and Environmental Sanitation Company (CETESB), the pollution control agency of the Environment Secretary of São Paulo state has recently been working to establish quality reference and intervention values for soils and groundwater, besides complying with the quality standards set by the law. These guiding values, defining the quality management in these media, have been adopted by CETESB, and were published in the report on the establishment of quality reference values and the intervention for soil and groundwater of São Paulo (CETESB 2001), as well as in the "Diário Oficial do Estado de São Paulo (2001)".

The indicative values that define intervention thresholds for groundwater were established as potability standards by the Ministry of Health, administrative rule n. 518, 25 March 2004, which revoked the rule of the Ministry of Health administration rule n. 1469, of December 29, 2000. An intervention value was established for the non-regulated substances, starting from the maximum acceptable concentration in soil in agricultural areas (APMax, Maximum Protection Area). Exceeding the values indicates the need for some kind of intervention in the area, due to the potential risk to human health (CETESB 2001). Table 5.2 presents the intervention values for soil and groundwater and those values from international institutions. Arsenic concentrations in agricultural soils have been strictly regulated in several countries, due to related possible health and environmental risks (Table 5.3).

The CONAMA resolution 420/2009 considers prevention of soil contamination necessary, aims at maintaining soil functionality and protection of surface and groundwater quality, and points out that contaminated areas can result in serious public health and environment risks. The resolution provides criteria and guiding values for soil quality for the presence of chemical substances and establishes guidelines for environmental management of contaminated areas (Table 5.4).

The law of the state of Connecticut provides As limits in soil with 10 mg kg^{-1} (De Capitani *et al.* 2002). The usual As range in soils is 0.1–40 mg kg^{-1}, and in vegetables 0.1–5 mg kg^{-1} (Brady 1983, in Rohde 2004). However, these values may be exceeded on both the low and high end of the spectrum (▶ 1). To evaluate the chemical quality of stream sediments in respect to arsenic, different governmental agencies in North America have adopted varying standards and values. These reference values result from the correlation between the adverse effects to aquatic organisms and the concentration of some chemical elements, usually the most toxic ones (CPRM 2004).

Table 5.2.   Soil As guide values (mg kg⁻¹) for São Paulo state with international values.

	São Paulo state					Holland	USA	
	Intervention values						Soil ingestion	
Reference value 3.5	Alert 15	Agri-culture APMax 25	Resi-dential 50	Industry 100		[1] 55	Res. 0.4	Industry 3.8
Germany			**Canada**				**England**	**France**
Trigger values (direct soil ingestion)								
Play-ground 25	House-hold 50	Park 125	Industry 140	Agric. 20	Househ. 30	Ind. 50	A B C D E 10 40 40 100 200	

*Based on risk to children: agriculture (agric.)/APMax;
Holland – [1]: multinationality;
England: A: home gardens;
B: playgrounds, parks, golf courses; France:
C: household (househ.),
D: park, E: industry (ind.).

Table 5.3.   Maximum As concentrations (mg kg⁻¹) in agricultural soils.

Concentration proposed in different countries									
Austria 1977	Poland 1993	Germany 1984	Russia 1988	UK 1987	USA 1988	Germany 1992	EU 2009	USA 1993	USA 1998
50	30	20	2	10	14	20	32	–	41

Kabata-Pendias and Pendias (2001), except for data referring to 1998 (Chaney *et al.* 1998).

Table 5.4.   Soil and groundwater As guide values.

Soil (mg kg⁻¹ dry weight)					Groundwater (μg L⁻¹)
		Investigation			
Quality reference	Prevention	Agricultural APMax	Residential	Industrial	Investigation
E	15	35	55	150	10*

E: defined by the State;
*potability standards of from chemical substances that pose health risks as defined in Ordinance n. 518/2004 of Brazilian Ministry of Health (CONAMA 2009).

The Brazilian Ministry of the Environment, through CONAMA, approved the resolution n. 344 (25 March 2004), published in the Diário Oficial da União on 7 May 2004, where the general guidelines and minimum procedures are established to evaluate the dredged river and marine sediments in Brazilian waters. Table 5.5 shows the classification of As levels of dredged waste.

Table 5.5.   Arsenic classification levels in dredged waste (CONAMA resolution n. 344/2004).

Pollutant (mg kg⁻¹)	Classification levels dredged material (in dry material unit)			
	Fresh water		Brackish and salt water	
Arsenic (As)	Level 1	Level 2	Level 3	Level 4
	5.9	17	8.2	70

Level 1 = limit below which there is low probability of adverse effects on biota.
Level 2 = limit above which there is probable adverse effect on biota.

The guiding values adopted in Table 5.5 are referenced in the following official Canadian and North-American publications:

- Canadian sediment quality guidelines for the protection of aquatic life. Environment Canada (2002)
- Incidence of adverse biological effects within ranges of chemical concentrations in marine and estuarine sediments (Long *et al.* 1995)

According to the Company of Mineral Resources (CPRM 2004), CETESB uses the evaluation criteria of the quality of sediments defined for Canada by the Canadian Council of Ministers of the Environment (CCME 2003) for São Paulo state. CCME has established standards for sediment quality based on the evaluation of the adverse effects of the most toxic metals in aquatic organisms (Table 5.6). Most terrestrial foods contain less than 1 μg g⁻¹ of dry weight arsenic (Möllerke *et al.* 2003); the levels in sea food are substantially higher with up to 80 μg g⁻¹ (▶ 1, 13).

The WHO has lowered the Acceptable Daily Intake (ADI) from 50 to 2 mg kg⁻¹ body weight, recommending more studies to be carried out to elucidate the nature of the arsenical compounds in food, particularly in sea food, where the levels of this element are usually higher (Lawrence 1986; Mantovani and Angelucci 1992, in Möllerke *et al.* 2003). The Brazilian Sanitation Surveillance Service of the Ministry of Health set a maximum level of 1.0 mg kg⁻¹ for As contamination in fish and fish products (Möllerke *et al.* 2003; Table 5.6). With regard to chemical contaminants in food, the Ministry of Health – Sanitation Surveillance Secretary published in the Diário Oficial da União of 28 August 1998, the Administrative Rule n. 685, of 27 August 1998, that set the "*general principles for establishing the maximum level of chemical contaminants in food*" and its annex: "*maximum tolerance limits for inorganic contaminants arsenic, copper, tin, lead, cadmium and mercury*". See Table 5.6 for the respective As limits.

In Spain, the law sets a maximum limit of 3 mg As kg⁻¹ for vegetables used in food, while the limit for certain As species is 3 μg L⁻¹ (Cunha 2006). The reference values for As levels in shrimp are usually around 650 μg kg⁻¹ (Yost *et al.* 1998 in Santos *et al.* 2003). The As limits in Chile are 1 μg g⁻¹ in fish and 2 μg g⁻¹ of inorganic arsenic in shellfish and crustaceous, as set by the Sanitation Regulation Agency (RSA; Vilches 2006). The WHO Expert Committee of Food Additives (JECFA/FAO/WHO) suggested a provisional value for the maximum tolerable weekly intake (PTWI) of 0.015 mg kg⁻¹ inorganic arsenic in body weight, corresponding to about 130 μg a day⁻¹ for a 60 kg person. This value does not apply to cases of acute exposure (FDA 2001 in De Capitani *et al.* 2002). Schachtschabel *et al.* (1984) report limit values for several chemical elements including arsenic, to be met by industry, besides values considered normal for As content in vegetables and in soil (Table 5.7).

Table 5.6.   Maximum As limits in food.

Type of food	As (mg kg^{-1})
Vegetable fats	0.1
Refined fats and emulsions	0.1
Hydrogenated fats	0.1
Sugars	1.0
Sweets and candy	1.0
Alcoholic and fermented beverages	0.1
Yeast- distilled alcoholic beverages	0.1
Cereals and products made from cereals	1.0
Ice cream	1.0
Eggs and egg products	1.0
Fluid milk, ready for consumption	0.1
Honey	1.0
Fish and fish products	1.0
Cocoa products and derivatives	1.0
Tea, coffee and derivatives	1.0

Admin. Rule n. 685/1998 (Ministry of Health 1998).

Table 5.7.   Arsenic limits and guiding values for drinking water, cereals, soils and potatoes, and normal levels for plants and soils.

Arsenic limits and guiding values				Normal As values	
Drinking water (mg L^{-1})	Cereals (mg kg^{-1}) wet weight	Soils (mg kg^{-1}) dry weight	Potatoes (mg kg^{-1}) wet weight	Plants (mg kg^{-1})	Soils (mg kg^{-1})
0.04	0.5	20	0.2	0.01–1	2–20

Note: the soils under study are from the former Federal Republic of Germany; the values for plants refer to leaves (Schachtschabel et al. 1984).

Table 5.8.   Human tolerance limits for As-compound exposure (Brasil 1978).

Agency	Description	Limit
Brazilian Ministry of Labor and Employment	Arsine (arsenamine) limit of tolerance (up to 48 h week^{-1})	0.04 mg m^{-3} (parts of vapor or gas per million parts of contaminated air) and 0.16 mg m^{-3}

In the human body, arsenic occurs at levels from 0.0005 to 0.032 μg g^{-1} (blood); levels around 0.01 μg g^{-1} are considered normal in urine, and normal levels in hair are below 1 μg g^{-1} (WHO 1981; in Pereira et al. 2002). According to Larini et al. (1997), the normal As content in urine is 10–100 μg L^{-1}; in blood 10–80 μg L^{-1}; in hair 0.5–2.1 μg g^{-1}; in finger nails 0.82–3.5 μg g^{-1}; and in toe nails 0.52–5.6 μg g^{-1} (▶ 1, 2).

The occupational As exposure generally occurs through the inhalation of particulates left behind by copper mining and foundry processes, semiconductors and glass manufacture, pesticides manufacture and use in wood and agriculture. Usually, the As exposure is added to the inhalation from other metals such as lead, silver, copper, antimony, chromium and gold (De Capitani et al. 2002; Table 5.8). The biological monitoring of the occupational As exposure is carried out through testing for these elements in urine or hair (▶ 2, 14).

Table 5.9. Regulations and guidelines for arsenic and As compounds.

Agency	Description	Information	Reference
	International guidelines		
IARC	Carcinogenicity classification	Group 1[a]	IARC (2004)
WHO	Air quality guidelines (unit risk[b])	1.5×10^{-3}	WHO (2000)
	Drinking water quality guidelines	0.01 mg/L[c]	WHO (2004)
	National guidelines		
	a) Air		
ACGIH	TLV (TWA)	0.01 mg m^{-3}	ACGIH (2004)
EPA	Hazardous air pollutant	yes	EPA (2004b), 42 USC 7412
NIOSH	REL (15-minute ceiling limit)[d]	0.002 mg m^{-3}	NIOSH (2005)
	IDLH[d]	5 mg m^{-3}	
OSHA	PEL (8-hour TWA) for industry in general	0.5 mg m^{-3}	OSHA (2005d), 29 CFR 1910.1000
OSHA	PEL (8-hour TWA) for industry in general	10 μg m^{-3}	OSHA (2005c), 29 CFR 1910.1018
	PEL (8-hour TWA) for construction industry	0.5 mg m^{-3}	OSHA (2005b), 29 CFR 1926.55
	PEL (8-hour TWA) for shipyard industry	0.5 mg m^{-3}	OSHA (2005a), 29 CFR 1915.1000
	b) Water		
EPA	Designated as hazardous substances in accordance with section 311b2 A of the Clean Water Act: arsenic pentoxide, arsenic trioxide, calcium arsenate and sodium arsenite	yes	EPA (2005d), 40 CFR 116.4
	Drinking water standards and health advisories DWEL	0.01 mg L^{-1}	EPA (2004a)
	National primary water standards a) MCLG	zero	EPA (2002a)
	MCL	0.01 mg L^{-1e}	
	Reportable quantities of hazardous substances designated pursuant to section 311 of the Clean Water Act: see above	1 pound	EPA (2005e), 40 CFR 117.3
EPA	Water quality criteria for human health consumption		EPA (2002b)
	Water + organisms	0.018 μg L^{-1f}	
	Organisms only	0.14 μg L^{-1f}	
	c) Food		
EPA	Tolerance for residues (dimethylarsinic acid)		EPA (2005i), 40 CFR 180.311
	Cotton (undelinted seed)	2.8 mg kg^{-1}	
	(Methanearsonic acid)		EPA (2005j), 40 CFR 180.289
	Cotton (undelinted seed)	0.7 mg kg^{-1}	
	Cotton, hulls	0.9 mg kg^{-1}	
	Fruit, citrus	0.35 mg kg^{-1}	
FDA	Bottled drinking water	10 μg L^{-1}	FDA (2005), 21 CFR 165.110
USDA	Non-synthetic substances prohibited for use in organic crop production	arsenic	USDA (2004), 7 CFR 205.602

(Continued)

Table 5.9.   Continued

Agency	Description	Information	Reference
	d) Other		
ACGIH	Carcinogenicity classification (biological exposure indices for inorganic arsenic plus methylated metabolites in urine at the end of the workweek) A1[g]	35 μg As L^{-1}	ACGIH (2004)
EPA	Carcinogenicity classification	group A[i]	IRIS (2007)
	Oral slope factor	1.5 mg kg^{-1} d^{-1}	
	Inhalation unit risk	4.3 ng m^{-3}	
	RfD	30 μg kg^{-1} d^{-1}	
	Superfund, emergency planning and community right-to-know		EPA (2005f), 40 CFR 302.4
	Designated CERCLA hazardous substance, reportable quantity		
	Arsenic not applicable [j]		
	Arsenic acid, arsenic pentoxide, arsenic trioxide, calcium arsenate, dimethylarsenic acid and sodium arsenite	1 pound	
	Effective date of toxic chemical release reporting for arsenic	01/01/1987	EPA (2005 h), 40 CFR 372.65
	Extremely hazardous substances, reportable quantity		EPA (2005 g), 40 CFR 355, App. A
	Arsenic pentoxide, calcium arsenate, and sodium arsenite	1 pound	
	Threshold planning quantities (arsenic pentoxide)	100/10.000 pounds	
	Calcium arsenate and sodium arsenite	500/10.000 pounds	
NTP	Carcinogenicity classification: known human carcinogen		NTP (2005)
	Toxicological Profile for Arsenic, Adapted 2010.		

[a]Group 1: carcinogenic to humans.
[b]Cancer risk estimates for lifetime exposure to a concentration of 1 μg/m^3.
[c]Provisional guideline value: as there is evidence of a hazard, but the available information on health effects is limited.
[d]NIOSH potential occupational carcinogen.
[e]MCL will become effective on 01/23/06.
[f]This criterion is based on carcinogenicity of 10^{-6} risk.
[g]A1: confirmed human carcinogen.
[h]A3: confirmed animal carcinogen with unknown relevance to humans.
[i]Group A: known human carcinogen.
[j]Indicates that no reportable quantity is being assigned to the generic or broad class.
ACGIH = American Conference of Governmental Industrial Hygienists;
CERCLA = Comprehensive Environmental Response, Compensation, and Liability Act;
CFR = Code of Federal Regulations;
DWEL = drinking water equivalent level;
EPA = Environmental Protection Agency;
FDA = Food and Drug Administration;
IARC = International Agency for Research on Cancer;
IDLH = immediately dangerous to life or health;
IRIS = Integrated Risk Information System;
MCL = maximum contaminant level;

(Continued)

Table 5.9. Continued

MCLG = maximum contaminant level goal;
NAS/NRC = National Academy of Sciences/National Research Council;
NIOSH = National Institute for Occupational Safety and Health;
NTP = National Toxicology Program;
OSHA = Occupational Safety and Health Administration;
PEL = permissible exposure limit;
REL = recommended exposure limit;
RfC = inhalation reference concentration;
RfD = oral reference dose;
TLV = threshold limit values;
TWA = time-weighted average;
USC = United States Code;
USDA = United States Department of Agriculture;
WHO = World Health Organization

In Brazil, the Regulation Act n. 7 (Occupational health medical control programme, revised and published in the Diário Oficial da União on 30 December 1994, through the Administrative Rule n. 24, of 29 December 1994) establishes that urine is the biological indicator of recent As exposure. The normal reference value is (VR1) 10 µg g^{-1} of creatinine, and the maximum permitted biological rate (IBMP2) is 50 µg g^{-1} of creatinine (De Capitani *et al.* 2002). Other environmental policy instruments should be linked to the establishment of standards, such as the evaluation of environmental impact, licensing and review and monitoring of activities that are actually or potentially polluting. This would allow the joint execution of an environmental research development program to study As effects on man and nature, and could effectively support implementation mechanisms.

In the United States, the government is responsible for creating regulations and recommendations aiming to protect the public health. These regulations are laws and standards to be followed. The agencies responsible for developing these regulations are the Environmental Protection Agency (EPA), the Occupational Safety and Health Administration (OSHA) and the Food and Drug Administration (FDA). Other agencies have the responsibility to make recommendations with regards to the toxic substances such as the National Institute for Occupational Safety and Health (NIOSH) and the Agency for Toxic Substances and Disease Registry (ATSDR). The Environmental Protection Agency (EPA) established a regulation that set the As standard in drinking water to 10 µg L^{-1}, aiming to protect consumers served by public water systems. This initiative was created to avoid long-term effects of chronic As exposure. This new standard for human consumption is valid from January 2006 and provides additional protection for Americans. In 2007, the ATSDR published the "Toxicological profile for arsenic" with limit value allowable for the various forms of chemical exposure, based on human exposure, relevance to public health and oral exposure. All established values recommended in the United States are presented in Table 5.9.

Arsenic compounds cannot be used alone or as constituents of preparations intended to prevent the proliferation of microorganisms, plants or animals on the hulls of boats, cages, floats, nets and any other appliances or equipment used for the cultivation of fish or shellfish or any appliances or equipment, partially or totally submerged, according to the directive EC 76/769/EEC of the European Environment Agency, dealing with restrictions on marketing and use of certain dangerous substances and compounds. This directive also prohibits their use in wood preservation. In this case, the prohibition does not apply to the solutions of inorganic salts of CCA (copper–chrome–arsenic) employed in industrial installations using vacuum or pressure to impregnate the solution into the wood. Furthermore, the member states of the European Union may authorize in their territory the use of preparations of DFA (dinitrophenol–fluoride–arsenic) for retreatment, in situ, of wooden poles already installed, with overhead cables. The As compounds, according to that directive can not be used alone or as constituents of preparations intended for use in the treatment of industrial waters, irrespective of their use. The evolution of environmental rules regulating the release of pollutants into water courses and in the atmosphere, and the disposal of waste on the ground, improved quality of life, welfare of the population and sustainable use of natural resources. However, such efforts have not as yet minimized the suffering of people exposed to toxic elements, especially the population of some regions.

REFERENCES

ACGIH: Arsenic. Threshold limit values for chemical substances and physical agents and biological exposure indices. Cincinnati, OH, 2004. *American Conference of Governmental Industrial Hygienists.*

ATSDR: Toxicological profile for arsenic. US Department of Health and Human Services, Agency for Toxic Substances and Disease Registry, Atlanta, USA., 2000. For latest updates, see: http://www.atsdr.cdc.gov/toxprofiles/tp.asp?id=22&tid=3.

ATSDR: Toxicological profile for Arsenic. Draft, Agency for Toxic Substances and Disease Registry, US Department of Health and Human Services, 2004. www.atsdr.cdc.gov/ToxProfiles/

Brady, N.C.: *Natureza e propriedade dos solos.* 6th ed, Freitas Bastos, Rio de Janeiro, 1983, p.647.

Brasil: Ministério do Trabalho e Emprego. Segurança e Trabalho (1978) http://www.mte.gov.br.

Brasil: Ministério da Saúde, Secretaria de Vigilância Sanitária. Portaria n 685, de 27 de agosto de 1998. Regulamento técnico. Princípios gerais para o estabelecimento de níveis máximos de contaminantes químicos em alimentos. Diário Oficial da União; Poder Executivo, de 28 de agosto de 1998.

Brasil: Ministério da Saúde. Portaria n. 518, de 25 de março de 2004. Ministério da Saúde, Secretaria de Vigilância em Saúde, Coordenação Geral de Vigilância em Saúde Ambiental, Brasília, Editora do Ministério da Saúde, 2005, p.28.

CETESB: Relatório de Qualidade das águas subterrâneas no Estado de São Paulo 1998–2000. Companhia de Tecnologia de Saneamento Ambiental, São Paulo: *Série Relatórios/Cetesb* 96 (ISSN 0103-4103), 2001.

Chaney, R.L., Angle, J.S., Baker, A.J.M. & Li, Y.M.: Method for phytomining of nickel, cobalt, and other metals from soil. U.S. Patent 5,711,784. Date issued: 27 Jan. 1998.

CONAMA: Conselho Nacional do Meio Ambiente. Resolução n 344, de 25 de março de 2004 – "Estabelece as diretrizes gerais e os procedimentos mínimos para a avaliação do material a ser dragado em águas jurisdicionais brasileiras, e dá outras providências". Publicada no DOU de 07/05/04 (2004) http://www.cprh.pe.gov.br/downloads/reso344.doc.

CONAMA: resolution n. 357 (17.03.2005), Conselho Nacional do Meio Ambiente: Dispõe sobre a classificação dos corpos de água e diretrizes ambientais para o seu enquadramento, bem como estabelece as condições e padrões de lançamento de efluentes, e dá outras providências." Data da legislação: 17/03/2005, Publicação DOU n° 053, de 18/03/2005, 2009, p. 58–63 http://www.mma.gov.br/conama/.

CPRM: Projeto APA Sul RMBH: geoquímica ambiental, mapas geoquímicos escala 1:225:000. Companhia de Pesquisa de Recursos Minerais. In: da Cunha, F. & Machado, G.J. (eds): *Semad/CPRM* 7 80, Belo Horizonte, 2004.

Cunha, A.P.: *Aspectos actuais da fitoterapia.* 2006.

De Capitani, E.M., Sakuma, A.M. & Tiglea, P.: Ecotoxicologia do arsênio e seus compostos. Salvador: Centro de Recursos Ambientais, *Série cadernos de referência ambiental,* 11 2002, p.129. ISBN: 8588595109

EEC: Directive 76/769/EEC, The Council of The European Communities 6 June 2010. http://www.ctei.gov.cn/document/20080514141211361885.pdf.

Environment Canada.: Canadian sediment quality guidelines for the protection of aquatic life. Canadian environmental quality guidelines – Summary Tables, 2002.

EPA: Arsenic in drinking water, treatment technologies. United States Environmental Protection Agency. Washington D.C, 1997.

EPA: Proposed revision to arsenic drinking water standard. US Environmental Protection Agency, 2000. http://water.epa.gov/lawsregs/sdwa/arsenic/regulations_pro-factsheet.cfm.

EPA: National primary drinking water regulations: arsenic and clarifications to compliance and new source contaminants monitoring. Fed. Register, 66:14:6975. United States Environmental Protection Agency, 2001.

EPA: National primary drinking water regulations. Washington, DC: Office of Ground Water and Drinking Water, US Environmental Protection Agency, 2002a. EPA816F02013; http://www.epa. gov/safewater/consumer/pdf/mcl.pdf.

EPA: National recommended water quality criteria. Washington, DC: Office of Water, Office of Science and Technology, US Environmental Protection Agency, 2002b. EPA822R02047. http://www.epa. gov/waterscience/criteria/wqctable/nrwqc-2004.pdf.

EPA: Minor clarification of the national primary drinking water regulation for arsenic. Federal Register, 40 CFR Part 141. US Environmental Protection Agency, 2003. http://www.epa.gov/fedrgstr/ EPA-WATER/2003/March/day-25/w7048.htm.

EPA: Capital costs of arsenic removal technologies: US EPA arsenic removal technology demonstration program Round 1. United States Environmental Protection Agency, 2004. http://www.epa.gov/ nrmrl/pubs/600r04201/600r04201.pdf.

EPA: Drinking water standards and health advisories. Washington, DC: Office of Water, US Environmental Protection Agency, 2004a. EPA822R04005. http://www.epa.gov/waterscience/drinking/ standards/dwstandards.pdf.

EPA: Hazardous air pollutants. Washington, DC: US Environmental Protection Agency, 2004b. United States Code. 42 USC 7412. http://www.epa.gov/ttnatw01/187polls.html.

EPA: Designated as hazardous substances in accordance with Section 311(b)(2)(A) of the Clean Water Act. US Environmental Protection Agency, 2005a. Code of Federal Regulations. 40 CFR 116.4.

EPA: Reportable quantities of hazardous substances designated pursuant to Section 311 of the Clean Water Act. US Environmental Protection Agency, 2005b. Code of Federal Regulations 40 CFR 117.3. http://www.epa.gov/region6/6en/w/sw/cwahzmat.pdf.

EPA: Superfund, emergency planning, and community right-to-know programs. Designation, reportable quantities, and notifications. Code of Federal Regulations. 40 CFR 302.4. U.S. Environmental Protection Agency, 2005c. http://www.epa.gov/osweroe1/lawsregs.htm.

EPA: Superfund, emergency planning, and community right-to-know programs. Extremely hazardous substances and their threshold planning quantities. US Environmental Protection Agency, 2005d. Code of Federal Regulations. 40 CFR 355, Appendix A. http://www.epa.gov/osweroe1/lawsregs.htm.

EPA: Superfund, emergency planning, and community right-to-know programs. Toxic chemical release reporting. US Environmental Protection Agency, 2005e. Code of Federal Regulations. 40 CFR 372.65. http://ecfr.gpoaccess.gov/cgi/t/text/text-idx?sid=27d0dad4dd3d4c1069aad205b798e315&c= ecfr&tpl=/ecfrbrowse/Title40/40tab_02.tpl.

EPA: Tolerances and exemptions from tolerances for pesticide chemicals in food. US Environmental Protection Agency, 2005f. Code of Federal Regulations. 40 CFR 180.311. http://ecfr.gpoaccess. gov/cgi/t/text/text-idx?sid=27d0dad4dd3d4c1069aad205b798e315&c=ecfr&tpl=/ecfrbrowse/ Title40/40tab_02.tpl.

EPA: Tolerances and exemptions from tolerances for pesticide chemicals in food. US Environmental Protection Agency, 2005g. Code of Federal Regulations. 40 CFR 180.289. http://ecfr.gpoaccess. gov/cgi/t/text/text-idx?sid=27d0dad4dd3d4c1069aad205b798e315&c=ecfr&tpl=/ecfrbrowse/ Title40/40tab_02.tpl.

EPA: Superfund, emergency planning, and community right-to-know programs. Toxic chemical release reporting. US Environmental Protection Agency, 2005h. Code of Federal Regulations. 40 CFR 372.65. http://www.epa.gov/epacfr40/chapt-I.info/chi-toc.htm.

EPA: Tolerances and exemptions from tolerances for pesticide chemicals in food. US Environmental Protection Agency, 2005i. Code of Federal Regulations. 40 CFR 180.311. http://www.epa.gov/ epacfr40/chapt-I.info/chi-toc.htm.

EPA: Tolerances and exemptions from tolerances for pesticide chemicals in food. US Environmental Protection Agency, 2005j. Code of Federal Regulations. 40 CFR 180.289. http://www.epa.gov/ epacfr40/chapt-I.info/chi-toc.htm.

EPA: Arsenic in drinking water, 2010. http://water.epa.gov/lawsregs/rulesregs/sdwa/arsenic/index.cfm.

FDA: Guidance document for arsenic in shellfish. Food and Drug Administration. http://www. cfsan.fda.gov/~frf/guid-as.htm. Cited in De Capitani et al. (2002) Ecotoxicologia do arsênio e seus compostos. Salvador: Centro de Recursos Ambientais. Cadernos de referência, 11 2001, p.130.

FDA: Beverages. Bottled water. Final Rule. Food and Drug Administration, 2005. Code of Federal Regulations. 21 CFR 165.110. Fed Regist 70:33694-33701. http://www.fda.gov/OHRMS/ DOCKETS/98fr/05-11406.pdf.

FUNASA: Portaria n 1.469/2000, de 29 de dezembro de 2000: aprova o controle e vigilância da qualidade da água para consumo humano e seu padrão de potabilidade. Brasília Fundação Nacional de Saúde 32, 2001.

IARC: Monographs on the evaluation of carcinogenic risks to humans. Some inorganic and organometallic compounds – arsenic and inorganic arsenic compounds, Volume 2. International Agency for Research on Cancer, 1973, p.181.

IARC: Monographs on the evaluation of carcinogenic risks to humans. Arsenic and arsenic compounds, Volume 23. International Agency for Research on Cancer, 1980, p.438.

IARC: Overall evaluations of carcinogenicity: An updating of IARC Monographs. Supplement 7: Arsenic and arsenic compounds (group 1), Volume1–42. International Agency for Research on Cancer, 1987, p.440.

IARC: Monographs on the evaluation of carcinogenic risks to humans. Some drinking water disinfectants and contaminants, including arsenic, Volume 84. International Agency for Research on Cancer, 2004a, p.512.

IARC: Overall evaluations of carcinogenicity to humans. as evaluated in IARC Monographs, Volumes 1–82. (at total of 900 agents, mixtures and exposures). Lyon, France: International Agency for Research on Cancer, 2004b. http://www-cie.iarc.fr/monoeval/crthall.html.

IARC: Overall evaluations of carcinogenicity to humans, as evaluated in IARC Monographs, Volumes 1–82. (at total of 900 agents, mixtures and exposures). Lyon, France: Internat Agency for Research on Cancer, 2006. http://mcgill.ca/files/cancerepi/IARC_Monographs.pdf.

IRIS: Arsenic. Integrated risk information system. Washington, DC: U.S. Environmental Protection Agency, 2007. http://www.epa.gov/iris/subst/index.html.

Kabata-Pendias, A. & Pendias, H.: *Trace elements in soils and plants. 3ʳᵈ ed.;* CRC Press, USA, 2000, p. 432.

Larini, L., Salgado, P.E.T. & Lepera, J.S.: Metais. In: Larini, L. (ed): *Toxicologia.* 3ʳᵈ ed.; Manole, São Paulo, 1997, pp.131–135.

Lawrence, J.F., Michalik, P., Tam, G. & Conacher, H.B.S.: Identification of arsenobetaine and arsenocholine in Canadian fish and shellfish by high performance liquid chromatography with atomic absorption detection and confirmation by fast atom bombardment mass spectrometry. *J. Agric. Food Chem.* 34:2 (1986), pp.315–319.

Long, E.R., Macdonald, D.D., Smith, S.L. & Calder, F.D.: Incidence of adverse biological effects within ranges of chemical concentrations in marine and estuarine sediments. *Environ. Manage.* 19:1 (1995), pp.81–97.

Mantovani, D.M.B. & Angelucci, E.: Avaliação do teor de arsênico em atum e sardinha. *Bol. SBCTA.* 26:1 (1992), pp.1–5.

Möllerke, R., Noll, I.B., Santo, M.A.B.E. & Norte, D.M.: Níveis de arsênio total como indicador biológico, na avaliação da qualidade do pescado (*Leporinus obtusidens* and *Pimelodus masculatus*) do Lago Guaíba em Porto Alegre RS-Brasil. *Rev. Inst. Adolfo Lutz* 62:2 (2003), pp.117–121.

NIOSH: Arsenic. NIOSH pocket guide to chemical hazards. Atlanta, GA, 2005. National Institute for Occupational Safety and Health, Centers for Disease Control and Prevention. http://www.cdc.gov/niosh/npg/npgsyn-a.html.

NTP: Report on carcinogens. 11ᵗʰ ed. Research Triangle Park, NC, 2005. U.S. Department of Health and Human Services, Public Health Service, National Toxicology Program. http://ntp.niehs.nih.gov/ntp/roc/toc11.html.

OSHA: Air contaminants. Occupational safety and health standards for shipyard employment. Occupational Safety and Health Administration, 2005a. Code of Federal Regulations. 29 CFR 1915.1000. http://www.osha.gov/comp-links.html.

OSHA: Gases, vapors, fumes, dusts, and mists. Safety and health regulations for construction. Occupational Safety and Health Administration, 2005b. Code of Federal Regulations. 29 CFR 1926.55, Appendix A. http://www.osha.gov/comp-links.html.

OSHA: Inorganic arsenic. Occupational safety and health standards. Occupational Safety and Health Administration, 2005c. Code of Federal Regulations. 29 CFR 1910.1018. http://www.osha.gov/complinks.html.

OSHA: Limits for air contaminants. Occupational safety and health standards. Washington, DC: Occupational Safety and Health Administration, 2005d. Code of Federal Regulations. 29 CFR 1910.1000. http://www.osha.gov/comp-links.html.

Pereira, S.F.P., Ferreira, S.L.C., Costa, A.C.S., Saraiva, A.C.F. & Silva, A.K.F.: Determinação espectofotométrica do arsênio em cabelo usando o método do dietilditiocarbamato de prata (SDDC) e trietanolamina/CH_3 como solvente. *Eclet. Quim.* 27 2002, p.14. São Paulo. http://www.cepis.org.pe/bvsacd/arsenico/simone.pdf.

Rohde, G.M.: *Geoquímica ambiental e estudos de impacto.* 2nd. ed. Signus Editora, São Paulo, 2004, p.157.

Santos, E.C.O., de Jesus, I.M., Brabo, E.d.S., Fayal, K.F., Sá Filho, G.C., de Oliveira Lima, M., Miranda, A.M.M., Mascarenhas, A.S., Canto de Sá, L.L., da Silva, A.P. & de Magaelhães Câmara, V.: Exposição ao mercúrio e ao arsênio em Estados da Amazônia: síntese dos estudos do Instituto Evandro Chagas/FUNASA. *Rev. Bras. Epidemiol.* 6:2 2003, pp.171–185.

Schachtschabel, P., Blume, H.P. Brümmer, G.W., Hartge, K.H., Schwertmann, U., Fischer, W.R., Renger, M. & Strebel, O.: *Lehrbuch der Bodenkunde.* Stuttgart, Enke. 12th ed: 1984, p.442.

Secretaria de Segurança e Saúde do Trabalho (1994). Portaria n 24, de 29 de dezembro de 1994. Aprova a norma NR 7 – Programa de Controle médico de Saúde Ocupacional. Diário Oficial [da] República Federativa do Brasil, Poder Executivo, Brasília, DF, 29 dez. 1994. Seção I, pp.1651–1659. http://www.fooddesign.com.br/arquivos/legislacao/nr_7_pcmso.pdf.

USDA: Nonsynthetic substances prohibited for use in organic crop production. US Department of Agriculture, 2004. Code of Federal Regulations 7 CFR 205.602. http://www.access.gpo.gov/nara/cfr/waisidx_04/7cfr205_04.html.

Valencia, A.C.: Arsênico, normativas y efectos em la salud. *XIII Congresso de Ingenieria Sanitária y Ambiental AIDIS* Antofagasta, Chile, 1999.

Viereck-Götte, L. & Ewers, U.: Grundlagen und Verfahren der Ableitung von Richtwerten. In: Matschullat, J., Tobschall, H.J. & Voigt, H.J. (eds): *Geochemie und Umwelt. Relevante Prozesse in Atmo-, Pedo- und Hydrosphäre.* Berlin: Springer, 1997, pp.245–264.

Vilches, S., Andrade, G., Muñoz, O. & Bastías, J.M.: Determinación del contenido de arsênico em productos marinos entregados por el Programa de Alimentación Escolar (PAE), de la Junta Nacional Escolar y Becas (JUNAEB), VII región, Chile. In: Bundschuh, J., Armienta, M.A., Bhattacharya, P., Matschullat, J., Birkle, P. & Rodríguez, R. (eds): *Natural arsenic in groundwaters of Latin America. Internat. Congr.* México City, 2006. 20–24 June 2006, p.83.

WHO: *Environmental health criteria: arsenic.* World Health Organization. Geneva, 1981.

WHO: *Air quality guidelines.* 2nd ed. Geneva, Switzerland: World Health Organization, 2000.

WHO: Bartram, J., Thyssen, N., Gowers, A., Pond, K. & Lack, T. (eds): *Water and health in Europa.* Regional Publications, 2002. European Series 93.

WHO: *Guidelines for drinking-water quality. 3rd ed.* Geneva, Switzerland: World Health Organization, 2004. http://www.who.int/water_sanitation_health/dwq/gdwq3/en/.

Yost, L.J., Schoof, R.A. & Aucoin, R.: Intake of inorganic arsenic in the North American diet. *Hum. Eco. Risk Assess* 4:1 (1998), pp.137–152.

Section II
Iron Quadrangle, ARSENEX project
and related perceptions

CHAPTER 6

History and socioeconomy – Iron Quadrangle

Isabel Meneses, Friedrich Ewald Renger & Eleonora Deschamps

Minas Gerais has been a key driver of the economy and development in Brazil, ever since its European discovery in the early 16th Century. Today the State occupies, after São Paulo, the second position as the richest, most developed and populated state in Brazil (IBGE 2010). Located to the south of the modern capital of Minas Gerais, Belo Horizonte, the Iron Quadrangle is particularly rich in mineral resources.

6.1 A BRIEF HISTORY OF THE IRON QUADRANGLE

The history of Minas Gerais can be traced back to the first European settlement at the Brazilian coast. At that time, the Portuguese Crown had but one interest: the mineral and natural resources necessary to support the Crown and its policy. An estimated 100 indigenous nations lived in the region at the time, however, very little is known about them. The first European expeditions arrived in the 16th Century, but a political system developed here only about 100 years later. The exploitation of the rich mineral resources of the province that would later become the State of Minas Gerais needed to be organized.

The large basins of the São Francisco and Jequitinhonha rivers attracted adventurers and explorers. Called "bandeirantes" (because they were structured like military formations), these pioneer explorers left important historical records, such as those from 1674 about the explorer Fernão Dias Paes Leme (1608–1681). Leme organized a seven-year expedition on behalf of the Crown in search of silver and emeralds, like those discovered by the Spanish in Peru and Mexico.

The first gold findings in the future state of Minas Gerais were reported during the late 17th Century in the das Velhas river valley, between the modern towns of Sabará and Lagoa Santa, attributed to the remnants of the "bandeira" of Fernão Dias, namely his son Garcia Rodrigues Paes (ca. 1650–1738) and his son-in-law, Manoel Borba Gato (1649–1718). Those discoveries unleashed the first "gold rush" in modern history and gave rise to the foundation of numerous settlements in central Minas Gerais during the first decades of the 18th Century: e.g., Sabará, Ouro Preto (Vila Rica), Caeté, Nova Lima, Mariana, Congonhas, Ouro Branco, São João del Rei, and Tiradentes.

Farming, not only for subsistence, and the opening of new roads followed the development and the steady influx of mainly European settlers, primarily from Portugal in the mid 18th Century (about 600,000). In the same period, an estimated 300,000 African slaves were brought to the area to work in the developing mines and on the farms. The largest portion of the wealth was exported to Portugal, and little was left for regional development. This led to constant friction between the settlers, São Paulo citizens ("paulistas"), and the Portuguese colonists. Between 1707 and 1709, these conflicts were known as the "Guerras dos Emboabas" (War of the foreigners, from Tupi language). As a consequence of the victory of the Portuguese party, an administrative unit was created, the Captaincy of "São Paulo e Minas de Ouro". However, the encounter of Europeans, natives and Africans led to an unprecedented miscegenation in the colony.

In 1720, a rebellion took place in Vila Rica, when local miners insurged against the newly ordered installation of a gold smelting house by the Crown. Felipe dos Santos, one of the leaders of the rebellion, was executed. The Portuguese government separated Minas Gerais from São Paulo to make control easier and created the Captaincy of Minas Gerais, with Vila Rica as the capital.

In 1789, the "Inconfidência Mineira" (insurrection of Minas Gerais) plotted the first attempt to make the State an independent republic. This rebellion was inspired by the American and French revolutions, known by many Brazilians who had studied in Europe. The attempt failed, and its leader, Joaquim José da Silva Xavier, alias Tiradentes (1746–1792), was sentenced to death and executed on the gallows in Rio de Janeiro. His body was quartered and pieces were exhibited along the road to Vila Rica.

At the end of the 18th Century, alluvial gold reserves were depleted, mainly because the deposits near the surface were exploited and no investments in technological development were made (Fig. 6.1). Mining methods were very simple: washing in river beds or near surface open cuts ("talho aberto") from which the gold-bearing Earth was washed from the slope of the mountains and then concentrated in huge tanks ("mondeos") near the river valleys. The unique mechanisation consisted of the so-called "rosario": a type of bucket chain driven by a water wheel used to dry the river beds (Fig. 6.2).

Population pressure resulted in the expansion of the boundaries of Minas Gerais into parts of Bahia in the Jequitinhonha valley, and into Goias; the Triangle region (Triângulo Mineiro) was politically added to the west of Minas Gerais in 1816. Agriculture developed and Minas became an important food supplier for Rio de Janeiro. The beginning of the 19th Century saw a rising demand for working capital to import more modern mining technology, not previously available in Brazil. This opened the door for British companies who then bought many inoperative mines to develop them later. During colonial times Au-amalgamation with mercury was not in use.

The famous Morro Velho mine in Nova Lima, today property of AngloGold Ashanti Mineração, is a good example of this history. Discovered in the 1720s (Lima 1901), it was worked for more than 100 years by local miners. In 1834, the mine was bought by a London-based company and worked until 2003 AD, with some excavations as deep as 2,500 m below surface. During this 169-year period, the mine produced a total of approximately 330 t of gold (1834–2003; Fig. 6.3), and was the most productive gold mine in Brazil (Pires *et al.* 1996).

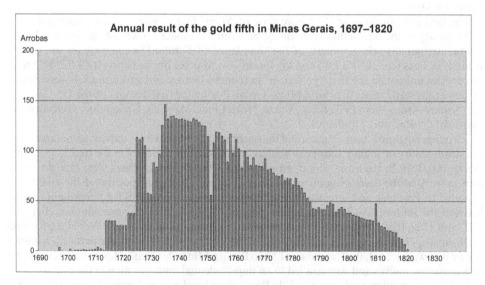

Figure 6.1.   Annual tax yield (the gold fifth) from Minas Gerais, 1697–1820. Au-production decreased steadily in Minas Gerais from 1735 AD onwards. The tax yield of the period 1700–1820 sums up to 7,526.4 arrobas or 110.5 t of gold (1 arroba = 14.688 kg), which corresponds to a production of 552.7 t of gold (certainly underestimated due to smuggling; Renger 2006).

Figure 6.2. Extraction of Au-bearing gravel from the das Velhas river during colonial times. The "rosario" (right) lowers the water level in the river bed and the gold-bearing gravel is dug manually by slaves (Renger 2005, from the collection of the Instituto de Estudos Brasileiros, USP, São Paulo, ca. 1780).

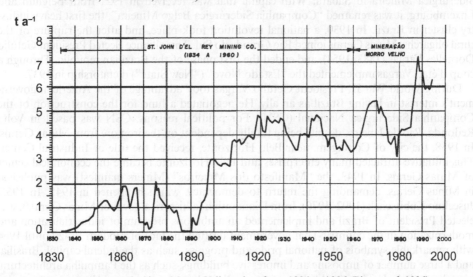

Figure 6.3. Annual Au production, Morro Velho mine, from 1834 to 1996 (Pires *et al.* 1996).

After 1808, with the relocation of the Portuguese court to Rio de Janeiro, greater political freedom and the dramatic reduction in Au production (Fig. 6.1) promoted radical changes in Minas Gerais. Geologists and other scientists came to the State. While they could not restore the gold reserves, their studies about the regional iron ore potential launched the basis for what would become the modern metal industry of Minas Gerais. The first industrial iron ore processing plant operations started in 1812, located near Congonhas in Minas Gerais.

Another unsuccessful revolution against the central government occurred in 1842 under the leadership of Teófilo Benedito Ottoni (1807–1869). In the mid 19th Century, Minas Gerais again became one of the most important provinces in Brazil. In addition to abundant sugar cane cultivation and livestock, coffee plantation started to give a new economic impulse, helping the State to obtain more political relevance.

Following the foundation of the Republic of Brazil in 1889, the power of the oligarchs grew. The so-called "coffee-with-milk" treaty between politicians from the states of São Paulo and of Minas Gerais basically divided the power, and the political influence in the country was shared between these two states. The agreement established that only politicians from these two states could take office as President of Brazil. From Minas Gerais, these were Afonso Augusto Moreira Pena (1847–1909), Venceslau Bráz Pereira Gomes (1868–1966), Delfim Moreira da Costa Ribeiro (1868–1920), Artur Bernardes (1875–1955), Juscelino Kubitscheck de Oliveira (see below) and Tancredo de Almeida Neves (see below). Due to the unfavorable geographical location of Ouro Preto (the old capital of Minas Gerais), it was decided to establish a new capital – Belo Horizonte, founded in 1897.

In 1908, further large iron ore reserves were found in the State, which soon attracted the attention of international investors. In 1910, the Itabira Iron Ore Company with British capital was the first to receive an operating license. At that time the expression "Quadrilátero Ferrífero" (Iron Quadrangle) was coined, as the geological structure of the region is formed by four megasynclines with the shape of a quadrangle and consists of very Fe-rich ore (mainly hematite), hosting also the Au deposits of Caeté, Mariana, Nova Lima, Ouro Preto, Sabará and Santa Bárbara (▶ 7.1).

In 1918, the American citizen Percival Farquhar (1864–1953) assumed the Company. In 1941, the Brazilian government reformed it to "Companhia Vale do Rio Doce" (CVRD). Today it is known as simply Vale, one of the world's three largest mining companies. In 1918, three Brazilian engineers, graduates of the School of Mines at Ouro Preto, founded the Companhia Siderúrgica Mineira in Sabará. With capital that was received in 1921 from Belgium and Luxembourg, it was renamed "Companhia Siderúrgica Belgo Mineira", the first heavy industry cluster in Brazil. In 1930, a national revolution took place, and after the rupture of the rural oligarchs, Minas Gerais joined Rio Grande do Sul. This alliance elected President Getúlio Dornelles Vargas (1883–1954), and ended the first phase of the Brazilian republic. Through a coup d'état, Vargas implemented the "Estado Novo" ("New State") dictatorship in 1937.

During World War II, President Getúlio Vargas took advantage of the American government's interest in having Brazil as an ally. He negotiated a fund for the construction of the Companhia Siderúrgica Nacional (CSN). For political reasons, CSN was based in Volta Redonda, Rio de Janeiro, despite being totally dependent on the iron ore from Minas Gerais. In 1938, the city of Contagem, near Belo Horizonte, received the title of Industrial Center. This initiative attracted many enterprises and Belo Horizonte became the economical center of Minas Gerais. In 1943, the "Manifesto dos Mineiros" (Miners manifest) was published in Minas Gerais, demanding the return to democracy, which happened in 1945. In 1955, Juscelino Kubitschek (1902–1976), from Diamantina (a historical town of Minas Gerais), was elected President of Brazil and implemented an ambitious program of industrialization and modernization in the country. That government drafted the face of modern Brazil until 1964 with remarkable symbols of national pride and progress, such as the federal capital, Brasília, and a large number of imposing and impressive buildings, such as the Pampulha architectural complex in Belo Horizonte. At the same time, these huge investments led to major national debts and high inflation. A new coup d'etat installed a military dictatorship that lasted until the end of 1984. Minas Gerais greatly benefited from those years, particularly by the development of metallurgical and mineral industries, and the installation of educational institutions. The next democratic phase, still in development, began with the indirect election of Tancredo Neves (1910–1985) as President of the Republic. The inauguration of this President Elect did not occur due to his sudden illness and death. Subsequently, the country was governed by the presidents José Sarney Costa (*1930), Fernando Collor de Mello (*1949), Itamar Franco (*1930), Fernando Henrique Cardoso (*1931), and Luiz Inácio Lula da Silva (*1945).

Today, Minas Gerais is the second most industrialized state in the country, after São Paulo, and is a major producer of iron ore, niobium, zinc and gold. Agriculture and livestock predominate in the south, the southeast and the west (Triângulo Mineiro). The high degree of development in these regions contrasts sharply with precarious conditions in the north and northeast of the State.

6.2 SOCIOECONOMY – NOVA LIMA AND SANTA BÁRBARA DISTRICTS

The described development is reflected in the socioeconomy of the region. Yet, and when compared to most European countries, conditions for the majority of the population still leave much to be desired. It is noteworthy, however, that important and positive steps were made between the years 1998 and 2007 (ARSENEX project activity period), as this section demonstrates. The ARSENEX project had been developed in the districts of Nova Lima and Santa Bárbara (▶ 7). The cities lie 22 and 105 km away from Belo Horizonte, respectively. Both belong to the meso-geographical metropolitan region of Belo Horizonte and to the micro-regions of Belo Horizonte (Nova Lima) and Itabira (Santa Bárbara).

Santa Bárbara district (684.71 km²). The district includes the town of Santa Bárbara, and was explored by the "bandeirantes" in their search for gold. Led by Antonio da Silva Bueno, they reached the banks of a creek and settled there in December 1704 for the exploitation of gold. In honor of the saint of that day they called the stream Santa Bárbara. The area around it soon developed into a settlement, called Santo Antonio do Ribeirão de Santa Bárbara, which emancipated from Mariana in 1839. The Santa Bárbara Mining Company was founded by British entrepreneurs in 1861. Gold mining in particular is still going on, leaving its marks on the landscape. The installation of the gold mines in the region contributed to the industrial activity. Despite the emphasis on the municipal economy, in addition to trade and services, forestry, agriculture and livestock were developed as means of subsistence.

Surveys conducted by the Brazilian Institute of Geography and Statistics (IBGE) from 1970 to 2009 (Table 6.1) show that the population of Santa Bárbara is predominantly urban. This urban population increased disproportionally – reaching 88% of the total (rural and urban) in 2000 – due to the better availability of jobs in urban centers, a typical feature (not only) of Brazil.

Table 6.1 shows a decrease in the population between 1991 and 2000. In 2004, however, a small increase in relation to 2000 occurred. This did not fully compensate the losses since the district Catas Altas emancipated from Santa Bárbara. According to IBGE, the district Santa

Table 6.1. Population residing in the city of Santa Bárbara (1970–2009).

Year	Urban inhabitants	Rural inhabitants	Total inhabitants
1970	9,223	7,023	16,246
1980	12,466	5,592	18,058
1991	20,969	4,962	25,931
2000	21,283	2,890	24,173
2004	–	–	25,239
2007	–	–	26,185
2009*	–	–	27,571

Note: IBGE (2010) – Demographic census data;
*2009 is an official estimate.

Bárbara currently consists of five municipalities: Santa Bárbara town, Barra Feliz, Brumal, Conceição do Rio Acima, and Florália. Table 6.2 shows the distribution of the economically active population. In 2000, about 62% of the population occupied the tertiary sector, which supports, overall, the industrial development.

According to the State Department of Finance (INDI 2005) tax collection increased steadily in the period 2000–2008, followed by a decrease in 2009 due to the global financial crisis (Table 6.3). Besides the Excise tax, Santa Bárbara received 2,914,971.25 R$ "Financial Compensation for Exploitation of Mineral Resources" (CFEM) in 2004 for the economic use of the mineral resources (DNPM 2004).

Nova Lima district (428.45 km²). The history of the town of Nova Lima has its origins, like Santa Bárbara, in "bandeirantes" from São Paulo, led by Domingos Rodrigues da Fonseca Leme and his brother Sebastião Pinheiro da Fonseca Raposo. In 1700, they discovered two gold-rich veins in the region; the beginning of the settlement, named originally Congonhas das Minas de Ouro. The district includes many old gold mines, today in the townships: Bela Fama, Cachaça, Vieira, Urubu, Gaia, Gabriela, Faria Garcês, Batista, Morro Velho. The economy then was based on the exploitation of the mineral deposits. The village of Congonhas das Minas de Ouro was elevated first to a parish, then to a district, subject to the municipality of Sabará. It was then named Congonhas de Sabará. In 1891, it was politically and administratively emancipated, and renamed Vila Nova de Lima, in honor of the resident Antônio Augusto de Lima (1859–1934), politician and historian, who ruled the state at the beginning of the Republican period. In 1923, it was named Nova Lima (INDI 2005).

Among the gold mining activities in Nova Lima, the Morro Velho Mine stands out, as it contributed to the consolidation of the urban center and the economy of the city. In 1834, the mine was acquired by the British company Saint John Del Rey Mining Company Limited, which explored the mine for a long time. The establishment of industries that supplied the British company with everything needed for its operation, such as candles, wooden boxes, gun powder, charcoal, etc, also contributed to this consolidation. In 1958, the British control

Table 6.2. Employees by economic sectors in 2000 and 2010 in Santa Bárbara district.

Sector	2000–2010
Agriculture, vegetable extraction, pesciculture	1,179–1,401
Industry	2,056–683
Trade	1,144–755
Services	4,015–1,033
Construction	n.d.–147
Total	**8,394–4,019**

Secretaria de Estado da Fazenda (2010), State Department of Finance; n.d.: no data.

Table 6.3. Santa Bárbara tax collection (2000–2009).

Year	ICMS Excise tax (R$)	Other (R$)	Total (R$)
2000	2,012,390	818,679	2,831,069
2001	2,457,805	938,422	3,396,227
2002	3,482,821	1,078,310	4,561,131
2003	4,259,952	1,315,734	5,575,686
2008	6,137,585	4,231,338	10,368,974
2009	3,194,584	3,667,085	6,861,670

Secretaria de Estado da Fazenda (2010), State Department of Finance.

of the property was transferred to the US-based Hannah Mining Company, which was interested only in the huge iron-ore deposits contained in its landholdings. So the gold properties were transferred to the newly created Mineraçao Morro Velho, controlled by Brazilian capital. Today, it belongs to the British-South African AngloGold Ashanti Mining Company. For centuries there has been a close relationship between the mining activities and the progress of the municipality, initially through the gold mining and since the 20th Century also through the development of iron ore deposits.

In this context, substantial tailing deposits accumulated in the districts of Mingu, Mina d'Água, Galo Velho, Galo Novo and Matadouro, which have been studied in the Project (▶ 11.2, 12). It should be noted that some families have been resettled mainly due to the expansion of these structures (▶ 9). Table 6.4 shows the development of the population in the urban and rural areas, reflecting again job generation in the industrial and services sectors.

Nova Lima is in the process of merging with the city of Belo Horizonte, as a result of the construction of middle-class and upper-class residential condominiums along the metropolitan axis of the roadways MG-030 and BR-040. Certainly, this occupation density will be reflected in a new form of distribution of the local residents. Considering the employment by economic sector (Table 6.5) in 2010, the services sector in general stood out, followed by the industrial sector, which includes transformation industry, mining and construction.

Table 6.6 shows the tax income from 2000 to 2009. It is important to mention that tax collection in Nova Lima is considerably higher than in Santa Bárbara (Table 6.3). In 2004, tax collection reached the total of R$ 11,377,089.51 through CFEM, representing the fourth largest tax collection of the State. This relative affluence explains the material differences between the two districts and highlights the uneven conditions in Minas Gerais, even in districts with a thriving mining industry.

Apart from these numbers it certainly deserves mentioning that Minas Gerais put a lot of emphasis on improving the school and educational system. Even more remote village schools

Table 6.4.   Resident population in Nova Lima district from 1970 to 2009.

Year	Urban inhabitants	Rural inhabitants	Total inhabitants
1970	27,377	6,651	33,992
1980	35,050	6,173	41,223
1991	44,038	8,362	52,400
2000	62,951	1,344	64,295
2007	–	–	72,207
2009*	–	–	76,608

Note: IBGE (2010) Demographic census from 1970 to 2009.
*2009 is an official estimate.

Table 6.5.   Employees by economic sectors in Nova Lima city in 2010.

Sectors	Number of employees
Agriculture, fishing	209
Industrial	6,837
Trade in goods	3,450
Services	15,185
Total	25,681

Secretaria de Estado da Fazenda (2000).

Table 6.6.   Nova Lima tax collection (2000 to 2009).

Year	ICMS (R$)	Other (R$)	Total (R$)
2000	7,999,013	3,621,631	11,620,644
2001	15,330,171	5,498,384	20,828,555
2002	10,654,270	5,665,180	16,319,450
2003	12,989,273	6,376,188	19,365,461
2008	63,268,140	29,332,598	92,600,739
2009	45,579,115	26,949,362	72,528,478

Secretaria de Estado da Fazenda (2003).

received a considerably improved infrastructure and continuous teacher training programmes over the past 15 years – resulting in increasing literacy rates.

Taking the presented socioeconomic numbers into account and looking at the historical development of the Iron Quadrangle region, there remains little doubt that particularly the last 30 years constituted a period of unprecedented growth not only in material wealth, but also in social wellbeing. Given the still rather large mineral resource potential of the region, one may not be surprised to see further development. The growing environmental awareness on many levels appears promising as well (▶ 8, 15, 17).

REFERENCES

DNPM: Compensação financeira pela exploração de recursos (CFEM). Departamento Nacional de Produção Mineral (2004) http: www.dnpm.gov.br.
IBGE: Censos demográficos 1970, 1980, 1991, 2000, 2004, 2007, and 2009. Instituto Brasileiro de Geografia E Estatística (2010) http://www.ibge.gov.br/home/estatistica/populacao/default_censo_2000.shtm.
INDI: www.indi.mg.gov.br/; Instituto de Desenvolvimento Integrado de Minas Gerais (Institute for Integrated Development of Minas Gerais), 2010.
Lima, A. de: Um município de ouro. *Revista do Arquivo Publico Mineiro* 6 (1901), pp.321–364.
Pires, A.S., Cunha, J.M.F. da, Lima e Fonseca TN de: Morro Velho. The story, events and achievements. Mineração Morro Velho Ltda. (ed), Nova Lima, 1996, p.205.
Renger, F.E.: Recursos minerais, mineração e siderurgia. In: Goulart, E.M.A. (ed): *Navegando o Rio das Velhas das Minas aos Gerais.* Instituto Guaicuy-SOS Rio das Velhas/Projeto Manuelzão/UFMG, Belo Horizonte, 2, 2005, pp.265–289.
Renger, F.E. O quinto do ouro no regime tributário nas Minas Gerais. *Revista do Arquivo Publico Mineiro* 42:2 (2006), pp.90–105.
Secretaria de Estado da Fazenda (2010) http://www.fazenda.mg.gov.br/index.jsp.

ADDITIONAL MATERIAL, INDIVIDUALLY:

Anglo Gold: Review of operations – Brazil. (2010) http://www.anglogold.com/subwebs/Information ForInvestors/AnnualReport04/report/review_of_year/brazil.htm.
Beirão, N.: History of Minas Gerais. In: Harder, B. (ed): *Minas Gerais Guide.* Brazil and Unibanco Guides. BEI Editora, São Paulo, 2006, ISBN 85-86518-65-4.
Braga, I.M.L.: Relatório Morro do Galo – Mineração Morro Velho Ltda, 2004, p.26.
Cantarin, C.: Em busca de ouro e da liberdade: uma breve história de Minas Gerais. In: Equipe BEI (eds) *Guia Ouro Preto e arredores.* Guia Caminho Real. BEI editora, São Paulo, 2006, ISBN 85-86518-70-0.
IBGE: Brasil em números. Brazil in figures. 14, 2006, p.349, Instituto Brasileiro de Geografia e Estatística, Rio de Janeiro.
Prefeitura Municipal de Santa Bárbara. Informações Municipais. http://www.santabarbara.mg.gov.br/.

CHAPTER 7

Physical aspects of the Iron Quadrangle

Katiane Almeida, Jörg Matschullat, Jaime Mello, Isabel Meneses & Zenilde Viola

The Iron Quadrangle (*Quadrilátero Ferrífero*) is a unique Pre-Cambrian structure between latitude 19°50′–20°30′ S and longitude 43°05′–44°30′ W (area ca. 7,000 km²). Apart from its mineral riches and its quality as a major iron, gold and gem producer, the relatively high morphological position moderates the tropical climate and allows for dense savanna (*Cerrado*) and Atlantic forests (*Mata Atlântica*) – wherever deforestation has not altered the landscape. Together, these two biomes account for more than 4,000 endemic plant species (Câmara and Murta 2007; Drummond *et al.* 2005; Gottsberger and Silberbauer-Gottsberger 2006). Sufficient precipitation sustains these biomes as well as agriculture and cattle breeding, and feeds the Doce and das Velhas rivers, two of the most important hydrological basins of Minas Gerais.

7.1 GEOLOGY – GEOMORPHOLOGY

While first maps date back into the 18th Century, the first serious geological work started with Wilhelm Ludwig von Eschwege (1777–1855) between 1822 and 1833 in the Ouro Preto area. The American Dr. Orville Adalbert Derby (1851–1915), then Director of the Brazilian Geological Survey, further explored and defined the geological formations of Minas Gerais between 1874 and 1912, in parallel to the French mineralogist Claude Henri Gorceix (1842–1919), founder of the School of Mines in Ouro Preto (today Federal University of Ouro Preto). Modern and systematic geological mapping and mineral exploration started after World War II with major input by the American geologist John Van Nostrand Dorr II (1910–1996), particularly on the iron and manganese deposits of the Iron Quadrangle (Dorr II 1969).

This physiographic and mineral province (Figure 7.1) extends in the north to the Serra do Curral mountain ridge at the southern rim of metropolitan Belo Horizonte. The Serra do Caraça forms its eastern boundary and the Serra do Ouro Branco defines the southern boundary. In the southwest, the Rola-Moça, Três Irmãos, Itatiaiuçu and Azul mountain ridges represent the farthest extension; and the Serra da Moeda defines its western limit. Located at the southern edge of the São Francisco craton, the Iron Quadrangle is composed by the following lithostratigraphic units (CPRM 2010; Figure 7.2):

- Archean crystalline basement, represented by portions of the Bação, Bonfim, Moeda, Maranhão, Belo Horizonte and Santa Bárbara metamorphic granite-gneiss complex complexes of Neoarchean age (2.7–2.85 Ga; Renger *et al.* 1994)
- Rio das Velhas Super Group, an Archaean greenstone belt. Its base consists of a volcano-sedimentary sequence (Quebra-Osso group), followed by meta-sedimentary and meta-volcanic schists (Nova Lima group), and a meta-sedimentary sequence (Maquiné group)
- Minas Super Group, formed by a sequence of metavolcanic and metasedimentary rocks (dated 2.58–2.1 Ga; Renger *et al.* 1994). It is subdivided into four groups: Caraça, Itabira, Piracicaba, and Sabará, characterized by quartzites, schists and phyllites of the Moeda formation, as well as by the phyllites from the Batatal formation. The Itabira group consists of banded iron formations (itabirites) from the Cauê formation and carbonates from the Gandarela formation. The Piracicaba group has five formations: Cercadinho, Fecho do Funil, Taboões, and Barreiro, overlain by the Sabará group.

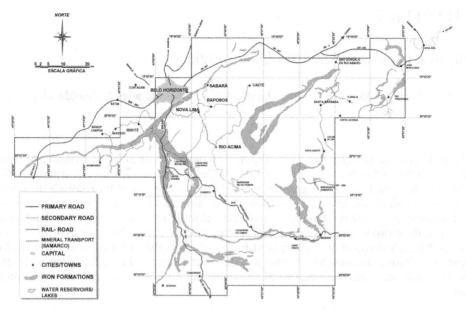

Figure 7.1.   Iron Quadrangle with iron deposits (large grey structures) and infrastructure (roads and major highways, towns and the capital Belo Horizonte (CERN 2007).

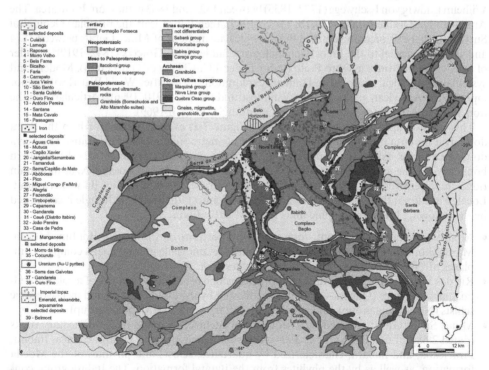

Figure 7.2.   Geological and mineral deposits map of the Iron Quadrangle (Dardenne and Schobbenhaus 2001).

The geomorphologic evolution of the Iron Quadrangle was shaped by a combination of structural conditions, lithology, epeirogenesis and climate, resulting in a distinct relief. The conditions determined the existence of suspended megasynclines and emptied anticlines, and structured crests. Ridges from 1,300 to 1,600 m are common as well as extensive erosion escarpments, many of which are constrained by faults, and hills with flat-bottomed valleys (1,000–1,100 m) in areas that are not controlled by the structure. The study region is located in the geomorphologic unit "dissected plateaus" (*planaltos dissecados*) of the centre-south and east of Minas Gerais. These plateaus were formed by granitic-gneissic rocks from which the hills and ridges with flat bottom valleys developed.

The **iron ore deposits** are hosted within the Cauê Formation, a typical Banded Iron Formation (BIF) of Lake Superior Type, which constitutes together with the Ganadarela formation the intermediate chemical member of the Minas Supergroup. The BIFs, locally known as itabirites, are metamorphosed and strongly oxidized. On the western border of the Iron Quadrangle, the Minas Supergroup is less deformed and exhibits many original sedimentary structures, such as cross bedding, graded bedding, ripple marks etc.. The metamorphism at this region reached the green schist facies. The iron ore bodies occur as discontinuous lenses of varied sizes and shapes within the itabirites. Two main types of iron ores are recognized: massive bodies of high-grade hematite ore (Fe > 64%) and intermediate-grade itabiritic ore (64% < Fe < 52%). The itabiritic ore is typically brittle and generally grades to hard itabirite with low Fe content. At the Águas Claras and Capão Xavier deposits, the host itabirite is dolomitic and the associated ore is very brittle, of very high-grade and low yield. The host itabirite is siliceous at the Tamanduá, Capitão do Mato and Pico deposits.

The origin of the high-grade iron ore has formerly been explained either by residual concentration of Fe-oxides after leaching of gangue minerals during Cenozoic weathering or by metasomatic processes during metamorphism. Recently, Rosière and Rios (2004) demonstrated a hydrothermal origin by fluid inclusion studies of those bodies formed in a 2-stage process, developing first a magnetite phase and afterwards a hematite phase during Transamazonian folding and metamorphism.

Gold is produced from numerous mines, mostly in the northern and southeastern parts of the district. The deposits are hosted in Archean or Paleoproterozoic carbonate or oxide-facies BIFs or iron-rich cherts in supracrustal sequences. Consequently, some of the old gold mines today are worked for iron (Lobato *et al.* 2001; Thorman *et al.* 2001). The four most important gold deposits in the Nova Lima Group are Morro Velho, Raposos, Cuiabá and São Bento, all hosted in Archean BIF and located in the districts of Nova Lima and Santa Barbara, respectively. Gold is associated with sulfide minerals and quartz-ankerite-dolomite in the Morro Velho mine. The dominant sulfides are pyrrhotite ($Fe_{1-x}S$), arsenopyrite (FeAsS), pyrite (FeS_2) and chalcopyrite ($CuFeS_2$). From 1962 to 1975, arsenic was a byproduct of the gold production, with a yield of approximately 100 t As_2O_3 a^{-1} (trioxide factory in Galo, Nova Lima). In the Raposos mines, gold and sulphides are located in a formation of siderite ($FeCO_3$). Pyrrhotite is the main sulphide in the ore, but arsenopyrite and pyrite are also present. In the Cuiabá mine, the ores contain mainly sulphides (pyrite and locally arsenopyrite), included in the banded iron formation (BIF). The gold deposit of São Bento is a sulphide replacement in iron formations of the Nova Lima group. Most of the gold is associated with arsenopyrite, pyrrhotite and to a lesser extent, pyrite (Vieira *et al.* 1991). Another important gold deposit is Passagem de Mariana, in the Ouro Preto district. Mariana is located in the vicinity of the contact between rocks of the Nova Lima group and the Minas Supergroup. Gold ore, very rich in pyrite, occurs in quartz veins and carbonates, Au-tourmalinite, phyllite and quartzite (Oliveira 1983; Vial 1988). Here, too, As trioxide production took place in the past, but accurate production data are not available.

The area between Itabira and Nova Era is the largest **emerald mining district** in Brazil, producing ca. 200 kg per month. In most cases, the beryllium needed for formation of the emerald deposits was provided by nearby pegmatites, which are consequently rich in beryllium and produce excellent gem-quality aquamarine and morganite (both $Be_3 Al_2(Si_6O_{18})$, the latter with some manganese; Cornejo and Bartorelli 2010; Morteani *et al.* 2000).

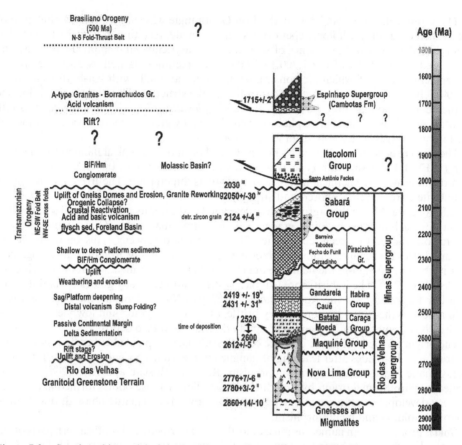

Figure 7.3.   Stratigraphic model of the Iron Quandrangle (Rosière *et al.* 2008).

7.2   SOILS OF THE IRON QUADRANGLE

A large diversity of soil types persists in Brazil, from highly weathered and deep-reaching soils to very shallow, poorly developed soils. Latosols (oxisols) predominate, occurring in about one third of the national territory (Ker 1995). Oxisols include deep, well drained and generally low natural fertility soil, with typical mineralogy at advanced stages of chemical weathering, especially with a predominance of kaolinite, gibbsite and ferric oxides, such as hematite and goethite in the clay fraction. In Minas Gerais, oxisols occupy a significant area along with ultisols (red clay soils), especially in flatter areas. In mountainous regions with strongly corrugated relief, the poorly developed shallow soils predominate. A peculiar soil type is the so-called canga, a crust of iron hydroxides with itabirite fragments, formed typically on outcrops of itabirite or massive hematite ore, which at many places also covers topographically lower strata of schists or phyllites.

Under these conditions, erosion is intense and does not allow developing deep profiles, although the climate is favorable to intense chemical weathering. The resulting soils are often shallow, poorly developed morphologically, but their mineralogy is typical of highly weathered soils and intensely leached, which gives them low natural fertility. Their chemical poverty contrasts with soils of temperate regions. It should also be noted that the soils are very old; the few available data indicating ages of many million years (Vasconselos *et al.* 1994).

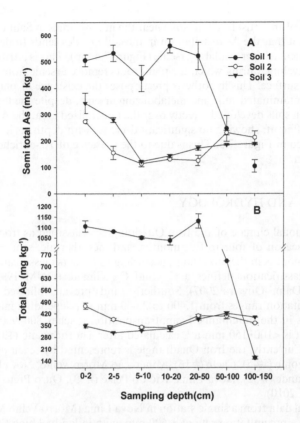

Figure 7.4. As distribution in soil profiles in Santa Bárbara. Top: available arsenic, Bottom: total arsenic with soil depth (Santana Filho 2005).

Ferriferous soils, the generic name for red soils with high iron content, develop from rocks rich in iron such as itabirites, magnetite-quartzite and conglomerates consisting of itabirite pebbles and ferruginous crusts (Camargo *et al.* 1982; Ker and Schaefer 1995; Oliveira *et al.* 1993). These soils can be classified into various taxonomic classes, mainly red oxisols/red latosols (also called "ferriferous"), inceptisols, entisols and petric plinthosols, which usually appear in areas of undulating hilly topography. The "ferriferous" latosols (>36% Fe_2O_3) are predominantly very rich in clay, sometimes with Fe and Mn nodules or concretions. They often occur in highland savanna regions at elevations between 1,000 and 1,300 m (Camargo 1982; Oliveira *et al.* 1993). These are low fertility soils with low cation exchange capacity (CEC), and with a low potential for agriculture. The inceptisols and entisols are shallow, and also of low natural fertility. Generally, they yield clayey loam and occur under highland savannas without agricultural use (Costa 2003). Despite the nomenclature and morphology, which presume development of incipient soil characteristics such as granular structure, CEC and mineralogy suggest highly weathered soils. The plintic plinthosols are soils of low fertility, with the presence of petroplinthite, features that restrict their agricultural use. In undeveloped areas near iron and gold mining, inceptisols and entisols predominate. Despite the low CEC, they have high capacity for phosphates and arsenate sorption, due to the remarkable presence of Fe oxides. These characteristics have environmental relevance, as they may affect As translocation from soils to waters, particularly in the As-rich soils near gold mining areas.

Recent studies of inseptisols (cambisols) near the mining areas in Santa Bárbara, Santana Filho (2005) noted that the As distribution in the soil profiles tends to decrease with depth (Figure 7.4), particularly for available arsenic (Figure 7.4 top). This is attributed to long-term biogeochemical cycling, during which the plant roots remove arsenic from deeper layers and deposit it on the surface. This hypothesis presupposes the existence of plant species capable of growing in contaminated soils and metabolizing arsenic, despite the high toxicity of this element (▶ 13). In soils developed directly over the mineralized area, the As distribution pattern is slightly different, showing no significant decrease with depth, as it is the case of soils 2 and 3 mentioned in Figure 7.3. This is due to the presence of As-enriched rock just below the soil C horizon.

7.3   CLIMATE AND HYDROLOGY

Climate. The regional climate of the Iron Quadrangle is classified as tropical and semi-humid with a dry season of four to five months; and mesothermic, semi-humid, again with four to five dry months in the mountains, according to Nimers classification (Nimer, 1989). The Koeppen classification defines a C_{wa} and C_{wb} climate (DNM 1992; INMET 2010; Mendonça and Danni-Oliveira 2007). Southerly winds prevail, followed by westerlies and easterlies. Precipitation ranges from 1,400 to 2,000 mm a^{-1}, depending mainly on elevation, with higher rates in the mountains. Evapotranspiration reaches 800–900 mm a^{-1}, and the hydrologic deficit is <100–150 mm a^{-1}, calculated after Thornthwaite (1955) for the hydrological balance. Currently, the Iron Quadrangle is represented by seven meteorological stations: Belo Horizonte (#45), Caeté (#16), Mineração Morro Velho, Nova Lima (#58), Nova Lima, Lagoa Grande (#52), Nova Lima, Rio do Peixe (#53), Ouro Preto (#73) and Sabará (#14) – (INMET 2010).

Meteorological data from a single station in Nova Lima (Morro Velho Mining, from 1855 to 2002) indicate a precipitation rate of 1,600 mm, with individual high (2,600 mm in 1906) and low (400 mm in 1963). A detailed observation of these meteorological data indicates distinct trends. From December to March, and in May, total monthly precipitation rates slightly decrease. There are particularly strong signals in December and February. April has a positive trend. No changes were observed in June to August, while the trend was negative again in September and November. October was positive (Pires et al. 1996). This trend should be the same for all the other stations in Minas Gerais, revealing impacts caused by the climate changes already perceived, particularly in the drier areas in the northern part of the state.

Hydrology. The area belongs to the sub-basins of the das Velhas and Piracicaba rivers (Figures 7.5 and 7.6), part of the São Francisco and Rio Doce hydrographic basins, respectively (ANA 2010). Rio das Velhas emerges in the Serra do Veloso, near Ouro Preto. Downstream from the city Pirapora the river flows into the São Francisco. Its major tributary in the project area is the Cardoso Creek, which became known as Água Suja (dirty water) stream, in the valley, where mining wastes with significant As levels were deposited (1930 to 1940). It was easier to access and cheaper to dump the mine wastes here along the tributaries. These deposits occupied the stream valley, namely Galo, Isolamento, Rezende and Madeira on the left bank, and the Matadouro and Fabrica de Balas deposits on the right banks. Part of the related risk for human health resulted from the sprawl of the Nova Lima urban area which forced direct contact of people with the deposits.

The confluence of the Socorro and Conceição rivers in the village of Barra Feliz forms the Santa Bárbara River that flows into the Piracicaba river at Bela Vista de Minas (Doce river basin). The Socorro river comes from the Espinhaço mountain range, which represents the division of the waters between the São Francisco and Doce river basins. The Conceição river has its source in the Capivari hills, a segment of the Espinhaço mountain range. Carrapato creek and the Caraça stream are tributaries of this river.

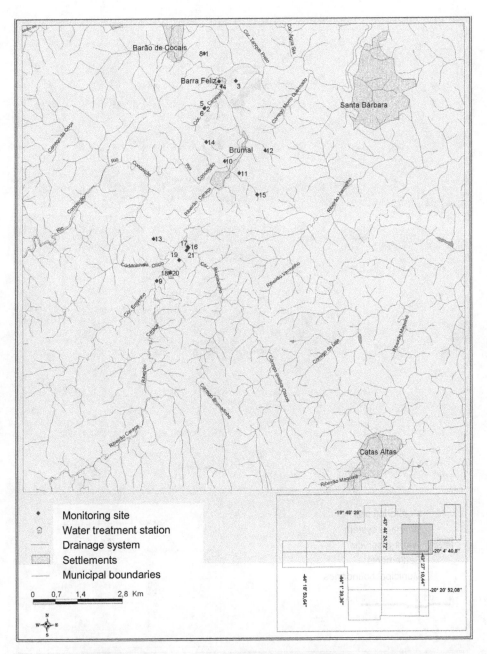

Figure 7.5.   Hydrographical map of the Santa Bárbara area. Small diamonds: monitoring site; house symbol: location of the water treatment plant between sites 18 and 20; shaded areas: inhabited community land; solid lines: drainage network; hatched line: district boundary (courtesy FEAM 2007).

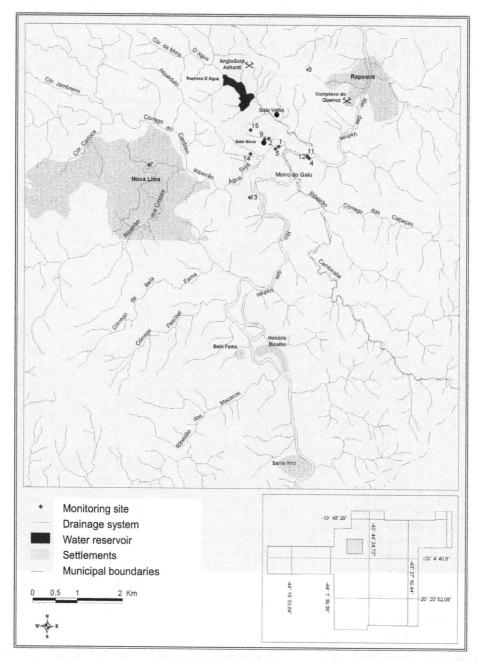

Figure 7.6.   Hydrographical map of the Nova Lima area. Small diamonds: Monitoring site; shaded areas: inhabited community land; solid blue area: water reservoir; solid lines: drainage network; hatched line: district boundary (courtesy FEAM 2007).

Table 7.1. Meteorological data (annual averages) Iron Quadrangle (EPAMIG 1982).

Parameter (unit)	Value range
Sunshine hours	2,000–2,500
Average air temperature (°C)	18–20
Minimum temperature (°C)	13–15
Maximum temperature (°C)	24–27
Relative humidity (RH%)	70–75
Evaporation (mm)	1,100–1,200
Precipitation (mm)	1,400–2,000

REFERENCES

ANA Agência Nacional das Águas, Brasília: Hydrological basin information, 2010. www.ana.gov.br.

Câmara, T. & Murta, R.: *Quadrilátero Ferrífero. Biodiversidade protegida.* Biblioteca da Pontefícia Universidade Católica de Minas Gerais, Belo Horizonte, 2007, p.200, (in Portuguese and English).

Camargo, M.N.: Proposição preliminar de conceituação de Latossolos ferríferos. In: Empresa Brasileira de Pesquisa Agropecuária (EMBRAPA), Serviço Nacional de Levantamento e Conservação de Solos (SNLCS). Conceituação sumária de algumas classes de solos recém reconhecidos nos levantamentos e estudos de correlação do SNLCS. Rio de Janeiro, Circular Técnica 1, 1982, pp. 29–31.

CERN: Consultoria Ambiental, Belo Horizonte, unpubl. and made available by FEAM 2007.

Cornejo, C. & Bartorelli, A.: *Minerals and precious stones of Brazil.* Solaris Cultural Publications, São Paulo, 2010, p.704.

Costa, S.A.D.: *Caracterização química, física e mineralógica e classificação de solos ricos em ferro do Quadrilátero Ferrífero.* M.Sc. thesis, Unpublished, Universidade Federal de Viçosa, Viçosa, Brazil, 2003, p.17.

CPRM: http://www.cprm.gov.br/estrada_real/geologia_estratigrafia.html, 2010.

Dardenne, M.A. & Schobbenhaus, C.: *Metalogênese do Brasil.* 1st ed., Volume 1, Universidade de Brasília, Brasília, Brazil, 2001, p.349.

Deschamps, E.M.: *Avaliação da contaminação humana e ambiental por arsênio e sua imobilização em óxidos de ferro e de manganês.* Doctoral thesis, Federal University of Minas Gerais, Belo Horizonte, Gerais, Brazil, 2003, p.139 + annex.

DNM: Normais climatológicas (1961–1990). Ministério da Agricultura e Reforma Agrária, Secretaria Nacional de Irrigação, Departamento Nacional de Meteorologia, Brasília, 1992, p.84.

Drummond, G.M., Soares Martins, C., Machado, A.B.M., Almeida Sebaio, F. & Antonioni, Y.: Biodiversidade em Minas Gerais. *Um atlas para sua conservação.* 2nd ed., Fundação Biodiversitas, Belo Horizonte, Brazil, 2005, p.222.

EPAMIG: Atlas climatológico do Estado de Minas Gerais. Empresa de pesquisa agropecuária de Minas Gerais. Instituto Nacional de Meteorologia/Universidade Federal de Viçosa, 1982.

Gottsberger, G. & Silberbauer-Gottsberger, I.: *Life in the cerrado, a South American tropical seasonal ecosystem. I. Origin, structure, dynamics and plant use.* Reta, Ulm, 2006, p. 277.

INMET: Maps and statistical station information CLINO 1961–1990, 2010. www.inmet.gov.br

Ker, J.C.: *Mineralogia, sorção e dessorção de fosfato, magnetização e elementos traços de Latossolos do Brasil.* PhD thesis, Universidade Federal de Viçosa, Gerais, Brazil, 1995, p.181.

Ker, J.C. & Schaefer, C.E.: Roteiro da excursão pedológica Viçosa-Sete Lagoas. In: *XXV Congresso Brasileiro de Ciência do Solo.* Departamento de Solos, Universidade Federal de Viçosa, Viçosa, Brazil, 1995.

Lobato, L.M., Ribeiro-Rodrigues, L.C. & Reis Vieira, F.W.: Brazil's premier gold province. Part II: Geology and genesis of gold deposits in the Archean Rio das Velhas greenstone belt, Quadrilátero Ferrifero. *Mineralium Deposita* 36:3–4 (2001), pp.249–277.

Mendonça, F. & Danni-Oliveira, I.M.: *Climatologia. Noções básicas e climas do Brasil.* Oficina do Texto, São Paulo, 2007, p.206.

Morteani, G., Preinfalk, C. & Horn, A.H.: Classification and mineralization potential of the pegmatites of the Eastern Brazilian Pegmatite Province. *Mineralium Deposita* 35:7 (2000), pp.638–655.

Nimer, E.: *Climatologia do Brasil.* 2nd ed., Fundação Instituto Brasileiro de Geografia e Estatística (IBGE), Rio de Janeiro, Brasil, 1989, p.412.

Oliveira, G.A.I., Clemente, L.C. & Vial, D.S.: Excursão à mina de ouro de Morro velho. In: *Anais do II Simpósio de Geologia de Minas Gerais,* Belo Horizonte, Soc. Bras. Geol., Núcleo Minas Gerais, Bol. 3, (1983), pp. 497–505.

Oliveira, J.B., Jacomine, P.K.T. & Camargo, M.N.: Classe gerais de solos do Brasil: Guia auxiliar para o seu reconhecimento. Jaboticabal, FUNEP, 1993, p.201.

Pires, A.S., da Cunha, J.M.F. & de Lima e Fonseca, T.N.: Morro Velho. *The story, events and achievements.* Mineração Morro Velho Ltda. (ed), Nova Lima, 1996, p.205.

Renger, F.E., Noce, C.M. & Romano, A.W. & Machado, N.: Evolução sedimentar do Supergrupo Minas: 500 Ma de registro geológico no Quadrilátero Ferrífero, Minas Gerais, Brasil. *Geonomos* 2 (1994), pp.1–11.

Rosière, C.A. & Rios, F.J.: The origin of hematite in high grade iron ores based on infrared microscopy and fluid inclusion studies: the example of the Conceição mine, Quadrilátero Ferrífero, Brazil. *Econ. Geol.* 99 (2004), pp.611–624.

Rosière, C.A., Spier, C.A., Rios, F.J. & Suckau, V.E.: The itabirite from the Quadrilátero Ferrífero and related high-grade ores: an overview. *Rev. Econ. Geol.* 15 (2008), pp.223–254.

Santana Filho, S.: *Distribuição de arsênio em solos e sedimentos e oxidação de materiais sulfetados de áreas de mineração de ouro do estado de Minas Gerais.* PhD thesis, Federal University of Viçosa, Minas Gerais, Brazil, 2005.

Thorman C.H., DeWitt E., Maron M.A.C. & Ladeira E.A.: Major Brazilian gold deposits – 1982 to 1999. *Mineralium Deposita* 36:3–4 (2001), pp.218–227.

Thornthwaite, C.W.: The water balance. *Climatol.* 8 (1955), pp.1–104.

Vasconselos, P.M., Renne, P.R., Brimhall, G.H. & Becker, T.A.: Direct dating of weathering phenomena by 40Ar/39Ar and K-Ar analysis of supergene K-Mn oxides. *Geochim. Cosmochim. Acta* 58:6 (1994), pp.1635–1665.

Vial, D.S.: Mina de ouro de Cuiabá, Q.F., Minas Gerais. In: Schobbenhaus, C. & Coelho, C.E.S. (eds): *Principais depósitos minerais do Brasil.* Volume 3, DNPM/CVRD, Brasilia, 1988, 413–419.

Vieira, F.W.R.: Textures and processes of hydrothermal alteration and mineralization in the Nova Lima Group, Minas Gerais, Brasil. In: Ladeira, E.A. (ed): *The economics, geology, geochemistry and genesis of gold deposits.* Brazil Gold'91; Balkema, Rotterdam, 1991, pp.319–325.

ADDITIONAL MATERIAL, INDIVIDUALLY:

Alkmim, F.F. & Marshak, S.: Transamazonian orogeny in the southern São Francisco cráton region, Minas Gerais: evidence for Paleoproterozoic collision and collapse in the Quadrilátero Ferrífero. *Precambrian Res.* 90:1–2 (1998), pp. 29–58.

Baltazar, O.F. & Silva, S.L.: Projeto Rio Das Velhas. Geological map of the Rio das Velhas Supergroup, 1:100.000; DNPM/CPRM, Brasília, 1996.

Baltazar, O.F. & Zucchetti, M.: Lithofacies associations and structural evolution of the Archean Rio das Velhas greenstone belt, Quadrilátero Ferrífero, Brazil: A review of the setting of gold deposits. *Ore Geol. Rev.* 32:3–4 (2005), pp. 471–499.

Dorr, II, J.v.N.: Physiographic, stratigraphic and structural development of Quadrilátero Ferrífero, Minas Gerais, Brazil. USGS. Prof. Paper, 641-A, 1969, p.110, Washington.

Leal, I.R., Tabarelli, M. & Cardoso da Silva, J.M.: *Ecologia e conservação da Caatinga.* 2nd ed., Editora Universitária UFPE, Recife, 2005, p.822.

Ottoni, C.: *Serra do Cipó – sempre viva.* Gráfica Editora Tavares, Pedro Leopoldo, Minas Gerais, 2008, p.327.

McGregor, G.R. & Nieuwolt, S.: *Tropical climatology.* 2nd ed., John Wiley & Sons, Chichester, 1998, p.339.

Sanches Ross, J.L.: Ecogeografia do Brasil. Subsídios para planejamento ambiental. Oficina de Textos, São Paulo, 2006, p.208.

Wallace, H.M.: The Moeda formation. *Boletim da Sociedade Brasileira de Geologia, São Paulo* 7:2 (1958), pp.59–60.

Zucchetti, M. & Baltazar, O.F.: Rio das Velhas greenstone belt lithofacies associations, Quadrilátero Ferrífero, Minas Gerais, Brazil. *31st Internat. Geol. Congr.*, Rio de Janeiro, CD-ROM, 2000.

CHAPTER 8

Project philosophy, history and development

Eleonora Deschamps & Jörg Matschullat

The idea for the ARSENEX project emerged in 1997. Some small studies from the State University of Campinas (UNICAMP) under the guidance of Prof. Dr. Bernardino Figueiredo, and preliminary work by Rawlins *et al.* (1997) with the British Geological Survey (BGS) suggested an As-related environmental problem near mining sites in the Nova Lima district. The latter work targeted (ground)water samples and aquatic plants (macrophytes). The results justified "a closer look" and posed the question as to whether any As contamination might have consequences for human well-being in the region.

From 1998 to 1999, UNICAMP developed a pilot-project in partnership with the Minas Gerais Environmental Agency (FEAM), with Dr. Jörg Matschullat, then Institute of Environmental Geochemistry, University of Heidelberg, and with the Baden-Württemberg State Health Agency in Stuttgart, Germany. The target area included Nova Lima, Santa Bárbara and Mariana (Figure 8.1), the latter representing the oldest industrial gold mining area in Brazil. Santa Bárbara district was first considered as a reference area for the project (baseline), as the local gold mining was supposedly less intense and more recent. This turned out to be an incorrect assumption (► 6).

The first major sampling campaign was carried out in April 1998, following a related workshop at UNICAMP with participation of the partner institutions. Samples were collected for a representative evaluation of As concentrations in most environmental compartments. Crustose lichen served as biomonitors for dust translocation and atmospheric

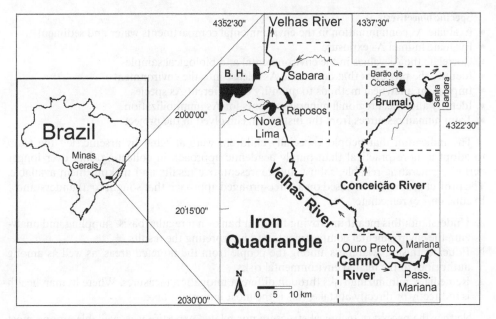

Figure 8.1.   The Iron Quadrangle in Minas Gerais, Brazil, with the key target areas Nova Lima, Santa Bárbara and Ouro Preto/Mariana (Matschullat *et al.* 2000).

deposition (▶ 10). Surface water was filtered in situ to differentiate between arsenic in the dissolved and the suspended phase (▶ 11) and the intermediate storage (reservoir) studied with surface sediment samples (▶ 12). Topsoil material (< 63 μm) was collected along streets and in private gardens, the latter to assess related accumulation, and to obtain an idea on possible soil-plant transfer (▶ 12). Plants, in particular edible ones from private gardens in the communities, were sampled to study soil-plant transfer and to assess the potential pollution pathway from soil via plant to human beings (▶ 13). The initial focus was directed to a first human biomonitoring (▶ 14). Urine from 126 public school children (8 to 12 years of age) was collected and processed. This first campaign revealed relevant As enrichment in all investigated environmental compartments (soil, sediments, water), and showed that a significant percentage of the investigated children were contaminated (Matschullat *et al.* 2000). These results justified further studies.

Based on the preliminary data, the project on As contamination was financed by the Ministry of the Environment and the Brazilian Environment Fund (FNMA) and varying support from German partner institutions (BMBF, DAAD, DFG) as of the year 2000. The team grew considerably and included the Fundação Estadual do Meio Ambiente (FEAM, an organ of the State Secretary for Environment and Sustainable Development), the Minas Gerais Sanitation Company (COPASA), the Ezequiel Dias Foundation (FUNED, an agency of the State Secretary of Health), the Engineering College of the Federal University of Minas Gerais (UFMG), the Water Chemistry Division of the University of Karlsruhe (ITC), and the Technical University Bergakademie Freiberg, Germany.

The fundamental goal of the study was a) to identify all possible As sources and their respective behaviour, b) to qualify and quantify As fluxes between the environmental compartments, c) to evaluate the environmental pressure and d) to evaluate the human exposure in the project area, as pre-conditions to e) justify possible proposals for mitigation action through governmental control. The objectives were defined as follows:

General objective:
- Meet and minimize As emissions into the environment and their impact on humans resulting from the decomposition of contaminant-laden materials.

Specific objectives:
- Evaluate As contamination in the environmental compartments water and sediment
- Evaluate human As exposure
- Correlate the As values in the environmental and biological samples
- Identify the processes that determine As release into the environment
- Implement analytical methods to quantify the different As species
- Identify and select regional mineral sorbents for As immobilization
- Train human resources from the institutions involved in the project

The indication that people, especially children, were affected by arsenic, led the team to adopt a more practical than purely academic approach in collecting data over longer periods, generating reliable, robust and representative results and making them available. The project proposal was based on a three-pronged approach that sought to a) understand – b) educate – c) remediate.

a. Understand: this meant improving the data bank on a regular basis, sampling and analyzing the environmental compartments and interpreting the results
b. Educate: raising awareness among the people from the targeted areas, as well as among authorities to the related environmental risks
c. Remediate: minimizing risks through different mitigation measures. Where human health is indirectly or directly at stake, this is the most important task.

Neither the necessary technical structure nor related expertise was available among most partners at the beginning of the project. Sampling campaigns were planned jointly with experienced members and with beginners, who were taught sampling, sample handling and processing procedures. This included strategy development, the preparation of the individual steps with a strong emphasis on quality control, including the proper storage and transport of the materials, so as to achieve correct and reliable analytical results adequate to the research objectives (▶ 10–14).

The standardization of the methodologies for As analysis among the participating groups began in the UFMG/DEMET, FUNED and COPASA laboratories, experimenting with internally developed methods and those learned in a training program held in the Instituto Adolf Lutz in Rio de Janeiro and in German laboratories. Thereafter, selected methods to analyse total As values in soils, sediments, plants, waters and effluents were implanted. This included result verification and quality control with round-robin analyses and the use of certified reference materials. It also included individual staff training both in Brazil and in Germany. This internal knowledge transfer was essential to develop and provide skills and the necessary infrastructure to successfully implement the project.

Many studies indicate drinking water as a major contributor to human As contamination. To control this ingestion pathway, the project sought to evaluate the most relevant natural sorbents for removal and immobilization of dissolved and particulate arsenic in water for human consumption (▶ 16). Iron oxides and hydroxides were identified due to their ability to sorb various anions, including aqueous arsenic (Huang 1994; Ladeira 1999; Pierce and Moore 1980; Schmidt 2001). Five natural sorbents from mines in Minas Gerais and a synthetic sorbent from Germany (GEH®) were selected. All of them were characterized in physical, chemical and mineral analyses in the laboratories, and their sorptive capacity was evaluated with standard techniques (▶ 16).

Based upon the first results, a suite of studies targeted the environmental compartments in more detail, and placed further emphasis on the human health issues. Clinical assessments were offered to the people, and further sampling and analysis of urine and human hair and fingernails was organized and performed (▶ 14). (Drinking) water, local vegetables, soils and sediments were studied in more detail (▶ 12, 13). The environmental and biological samples data, jointly with the data collected during the environmental perception studies (▶ 9), helped define the structure of "Workshops for environmental education and health". These events included lectures, movies, games and written activities with didactic material as well as other events suggested by the community (▶ 15). During the workshops, ways of dealing with the sources of As contamination were defined and proposals for corrective measures were made (▶ 15). An effective and dedicated longer-term involvement of the community emerged as a *"conditio sine qua non"* for any attempt in successfully reducing As-related risks in the region.

While the dedicated work of scientists and laboratory technicians is a prerequisite in obtaining reliable data, the dissemination of scientific results to the target audience is also essential for a project to become a success. It is well known that successful scientific communication is the key to changing attitudes. Only after understanding the implications of a given situation and the consequences of personal actions and behaviors, will citizens be willing to change or adapt to different situations.

During the sampling campaigns, scientists and laboratory technicians were invited to participate in these workshops and were introduced to the people from the target community. This personalization of the academic world helped to bridge the gap between the local professionals, the lay people and the members of the project team. This approach certainly made the acceptance of the suggested adaptation strategies easier.

A total of 23 workshops were held alternatively between the two districts Santa Bárbara and Nova Lima. Instruction and education materials, such as leaflets and booklets were used and distributed in the workshops and other activities held for environmental education. The project was also announced in the media, making the As-related issue known to in the target regions. In addition, presentations and discussions of project results took place in national and international scientific forums, to foster the exchange of information and the consequent suggestions of new approaches to the issue. With its formal end in 2007, a book was published in Brazil to further disseminate the experience and to help others to meet related challenges (Deschamps and Matschullat 2007).

REFERENCES

Deschamps, E. & Matschullat, J. (eds): *Arsênio antropogênico e natural. Um estudo em regiões do Quadrilátero Ferrífero*. Fundação Estadual do Meio Ambiente, Belo Horizonte, 2007, p.330, ISBN 978-85-61029-00-5.

Huang, Y.C.: Arsenic distribution in soils. In: Nriagu, J.O. (ed): Arsenic in the environment. *Advances in Environmental Science and Technology*, Volume 26, I. Wiley, New York, 1994, pp.18–49.

Ladeira, A.C.Q.: *Utilização de solos e minerais para imobilização de arsênio e mecanismo de adsorção*. PhD thesis in Metallurgical Engineering, Federal University of Minas Gerais, Belo Horizonte, Brazil, 1999.

Matschullat, J., Borba, R.P., Deschamps, E., Figueiredo, B.R., Gabrio, T. & Schwenk, M.: Human and environmental contamination in the Iron Quadrangle, Brazil. *Appl. Geochem.* 15:2 (2000), pp.181–190.

Pierce, M.L. & Moore, C.B.: Adsorption of arsenite on amorphous iron hydroxide from dilute aqueous solution. *Environ. Sci. Technol.* 14:2 (1980), pp.214–216.

Rawlins, B.G., Williams, T.M., Breward, N., Ferpozzi, L., Figueiredo, B.F. & Borba, R.P.: Preliminary investigation of mining-related arsenic contamination in the provinces of Mendoza and San Juan (Argentina) and Minas Gerais State (Brazil). In: British Geological Survey Technical Report WC/97/60, Kenworth, UK, 1997, p.25.

Schmidt, H.: *Arsenbelastung von Oberflächenproben subtropischer Böden aus dem Eisernen Viereck, Minas Gerais, Brasilien*. Unpubl. M.Sc. thesis, TU Bergakademie Freiberg, IÖZ, 2001, p.118.

CHAPTER 9

Environment and health perception

Adriano Tostes & Eleonora Deschamps

9.1 OBJECTIVES

Environmental perception studies provide researchers, public agents and educators with a better understanding of values, beliefs, attitudes and behaviours of a given population. Additional issues, such as options to be taken, e.g., to implement projects that promote economic development or that aim at environmental protection, can be on the agenda. From a community perspective, such studies are used to define priorities for investment and governmental action. Since the 1970's, communities have increasingly taken responsibility in managing and implementing actions. These actions would otherwise not be accomplished, due to the complexity of environmental challenges and difficulties the government experiences in responding to public demands. Not surprisingly, many international and national multilateral organizations have been given preference to projects that involve the direct participation of communities in solving their problems. The assumption is made that only the local community can identify the issues that affect them and put forward the best solutions along with their commitment to the proposed actions (Jacobi 1996).

It was this understanding that led the participating institutions in the ARSENEX project to complete a perception study. Social, cultural, health and environmental aspects that especially focus on As-related risks in the selected target-communities were analyzed. The resulting knowledge is considered essential in effectively supporting the public agents' action in the project area. Besides the technical-scientific knowledge, which is the basis for recommendations for action in the study area, the research on environmental perception sought to understand those parameters that influence the communities in facing their environmental challenges.

9.2 METHODOLOGY

This study was based on interviews. Its questions were selected according to the project objectives. To implement an environmental education project in the study region (▶ 15) successfully, most aspects of the social reality were included in this study. The targeted population lives under direct risk of exposure to As-contaminated sites. The survey was conducted in the townships of Barra Feliz and Brumal (Santa Bárbara district), and in the townships of Galo Novo and Galo Velho, in Mingu de Cima and de Baixo, in Matadouro and Isolamento, and in Mina d'Água (Nova Lima district).

9.2.1 Research instruments

This study included public opinion survey techniques. A semi-structured questionnaire was used for data collection with closed precodified questions, and open/free questions. The work was carried out by trained interviewers (Norris and Field 1993). The intention was to contribute effectively to the environmental action plans for the target communities, providing research results that would optimize the mitigation efforts jointly developed by the participating government institutions and the civil society.

The questionnaire concentrated on environmental and health aspects directly related to the main project focus, such as environmental exposure and As contamination in the communities. In addition to socio-demographic issues, which characterize these communities, the questionnaire approached subjects previously identified by the researchers as relevant, particularly those related to the As exposure and its contamination forms and the effects on human health. The questionnaire included eleven topics:

1. *Locality characterization* – information about county, district, etc.
2. *Family structure* – data on the respondent and her/his family, such as time living in the place, number of children, activities and leisure places, etc.
3. *Infrastructure* – data about the house, electricity supply, drinking water origin, waste destination, sewage, etc.
4. *Job* – profession related data, particularly whether the respondent works in mining or metallurgy, labour safety, prevention against contaminants, etc.
5. *Food health* – data on dietary habits (consumption of fruits and vegetables and their origin), food hygiene, fish consumption, etc.
6. *Knowledge about arsenic* – information that verifies the level of the respondent's knowledge about arsenic, its contamination forms and effects and also evaluates the level of the community perception related to development/environment/health and environment preservation.
7. *Diseases* – identification of symptoms related to arsenic contamination, such as cancer, habits related to alcohol and tobacco consumption, etc.
8. *Environmental perception* – knowledge about regional environmental issues, such as water, soil, and air pollution, etc.
9. *Involvement with environmental issues* – degree of respondent's willingness to engage in activities that protect the environment.
10. *Public bodies* – evaluate the official organs action in the environment area and related issues in the region.
11. *Personal identification* – data on name, address, age, educational level, etc.

9.2.2 Work development

Six steps were developed as follows:

1. **Research preparation**. Technical meetings were held with the project member institutions to define the general scope and to raise the main issues to be studied. The questionnaire was formulated and the team trained to administer the questionnaire.
2. **Field research**. Fieldwork started in the selected communities when the development of the questionnaire was finished.
3. **Analysis and control of the responses, encoding of the questionnaires and processing**. After completion of the fieldwork, the team gathered, read the questionnaires and coded the responses to the open questions. Answers were pre-coded for the closed questions. This work triggered a "book" of codes with all categories of answers for each question in the questionnaire. This code book enabled the subsequent coding of the questionnaire answers (post-codification). This codified answer sheet is the "mirror" of the research database.
4. **Assembly and consistency test of the database**. Once the encoded answers were transferred to the answer sheet, the data were entered onto a spreadsheet. SPSS for Windows (Statistical Package for Social Science SPSS BASE 10.000; 1999) was used. Finally, the work was checked for possible misspellings and/or inconsistency between the responses.
5. **Data Pre-analysis**. After checking database consistency, a preliminary study of the data started, and the responses readied for the final analysis. This consisted of a descriptive evaluation of the questions and their respective answers (frequency distribution), treated as variables. This work is highly relevant, as it indicates the form and contents of the variables

to be analyzed. Then, questions and answers were grouped into categories (re-coding), to obtain incidence rates of related diseases, for example. In this step an analysis grid was created, to identify the variables that would need to be processed in a descriptive form and that would require multivariate analysis. An analysis guideline was developed to identify the independent, the dependent and the intervening (or control) variables.

6. **Data Analysis and final report**. The final report identifies the problems and the potential solutions. Besides characterizing and describing the residents' perception of the reality in the communities, it identifies issues and strategies that could be considered for the environmental education workshops. Its purpose was to mobilize the communities to participate effectively in the improvement of their quality of life.

9.3 DISCUSSION AND ANALYSIS OF RESULTS AND PRODUCTS

The study was based on intentional sampling, obtained from populations that were directly exposed to potentially contaminated sites. This qualitative scope was of great value for designing the environmental education workshops for diverse communities (▶ 15). The analysis of the questionnaires allowed not only the identification of common issues for communities, but also specific differences. This can determine the success of public actions and policies to solve the key problems of the community. Table 9.1 quantifies the fieldwork per township, with the number of questionnaires; 343 in total.

9.3.1 *Socio-demographic characteristics*

Access to public services. Such access is a right; yet related availability often defines the level of local development, as the services contribute to improving life conditions and environmental quality. Urban solid waste may be used as an example, given the relevance of effective waste management – collection, treatment, destination – for the sanitation conditions of the residents. The vast majority (92.4%) of the respondents said they had waste collection service at home. However, 23% report the existence of alternative practices, such as waste burning.

Table 9.2 indicates that the population of Santa Bárbara, besides having a lower coverage of collection service, burns more waste – reaching up to 45.1% in Brumal – than other areas. Burning waste is detrimental to the environment and should be avoided, either by increasing residential waste collection services or environmental education. As well as posing a risk for human health, residue burning may cause fires, a problem mentioned by the interviewees (see below).

Table 9.1. Local questionnaire result statistics (sampling in 2002).

District	Township	Cases	Percentage	Valid %	Accum. %
Santa Bárbara	Barra Feliz	90	54.2	54.2	54.2
	Brumal	71	42.8	42.8	97.0
	Santa Bárbara city	5.0	3.0	3.0	100.0
	Total	**166**	**100.0**	**100.0**	-
Nova Lima	Nova Lima city	16	9.0	9.0	9.0
	Galo Novo/Velho	43	24.3	24.3	33.3
	Mingu de Cima/ de Baixo	47	26.6	26.6	59.9
	Matadouro/Isolamento	38	21.5	21.5	81.4
	Mina d'Água	33	18.6	18.6	100.0
	Total	**177**	**100.0**	**100.0**	-

Fewer respondants indicated they had access to sewage services: only 32.4% [= (n Santa Bárbara + n Nova Lima)/343] confirmed that their homes were connected to the sewage network. Most, 52.8% [= (n Santa Bárbara + n Nova Lima)/343], informed that their sewage is released into the river; 2.9% [= (n Santa Bárbara + n Nova Lima)/343] said that their sewage is discarded in the open, or in dams, 2.3% [= (n Santa Bárbara + n Nova Lima)/343]. Around 9% [= (n Santa Bárbara + n Nova Lima)/343] of the respondents said they had septic tanks in their homes.

This analysis by district and township indicated relevant differences: only 12% of homes in Santa Bárbara had this service. The townships of Barra Feliz and Brumal had the lowest coverage in Santa Bárbara. There, 76.7% and 73.2% respectively of the respondents said that sewage in their homes is discarded directly into the rivers.

Although larger, Nova Lima also has a relatively low percentage of 51.4% access. The negative highlight is the district of Galo Velho/Galo Novo, where only 30.2% of respondents stated that their homes were connected to the sewage network, and 60.5% discard their waste directly into the local rivers and creeks.

Many respondents use one or more drinking water supply sources (Table 9.3). Important differences were noted between the cities and the respective districts. The research reports that most respondents in Nova Lima are supplied by the Companhia de Saneamento de Minas Gerais (COPASA), which supplies 97.2% of the homes. In the Santa Bárbara district, access decreases to 19.9%.

People in the communities under study usually relied on alternative sources of supply (untreated water), with all implications for human health. Of the total respondents from Santa Bárbara, 80.7% used water from the springs; in Brumal this percentage reached 95.8%.

Table 9.2.   Waste destination per city and their respective districts.

Waste Destination

District	Township	Collection by truck		Buried		Burnt		Reused		Total
		n	%	n	%	n	%	n	%	n
Santa Bárbara	Barra Feliz	81	90.0	2	2.2	39	32.2	4	4.4	90
	Brumal	61	85.9	3	4.2	32	45.1	16	22.5	71
	Santa Bárbara city	5	100.0	5	3.0	2	40.0	1	20.0	5
	Total	**147**	**88.6**	–	–	**63**	**38.0**	**21**	**12.7**	**166**
Nova Lima	Nova Lima city	16	100.0	–	–	–	–	–	–	16
	Galo Novo/Velho	42	97.7	–	–	6	14.0	1	2.3	43
	Mingu de Cima/ de Baixo	46	100.0	–	–	2	4.3	–	–	46
	Matadouro/Isolamento	35	92.1	–	–	8	21.1	1	2.6	38
	Mina d'Água	33	100.0	–	–	–	–	–	–	33
	Total	**172**	**97.7**	–	–	**16**	**9.1**	**2**	**1.1**	**176**

Table 9.3.   Water supply in the Santa Bárbara and Nova Lima districts.

Sources

District		COPASA/SAAE	Well*	River/spring	Total
Santa Bárbara	n	33	33	134	166
	%	19.9	19.9	80.7	100.0
Nova Lima	n	172	2	19	177
	%	97.2	1.1	10.7	100.0

*artificial well.

In Nova Lima, despite the highest access rate to this public service in all districts, only 44.2% in Galo Novo/Galo Velho reported that they used water from the springs for domestic consumption. The situation is similar in most townships of Santa Bárbara and Nova Lima with respect to waste water treatment (Table 9.4). This is an issue, since traditional home methods of water treatment are not able to eliminate particulates, which is important when water is As contaminated. (▶ 11, 12).

The numbers in Tables 9.3 and 9.4 are of concern, as non-treated water supplies may not be potable and therefore, may transmit several diseases (▶ 14). Thus, the study suggests socio-educational measures be taken in order to reduce the health risks in these communities (▶ 15).

a) Local conditions and housing. Although most respondents (79%) lived in places with paved streets and/or sidewalks, some communities do not provide these amenities. In Nova Lima, a town with a higher rate of urban development, only 13% of respondents reported living in houses on unpaved streets. This number increased to 27.7% in Santa Bárbara.

There was little variation in housing construction, by township or by district. Most respondents (98.8%) reported living in brick homes with cement or rock floors. Despite these data, it is noteworthy that most people live in typical rural houses; many with flower and vegetable gardens, terraces, orchards and livestock.

b) Experience in the community. Most respondents, around 80%, have lived in the communities with their families for more than ten years; 53% for more than 20 years. This certainly influences people's involvement with their environment. All residents know the local rivers, streams and springs, although sometimes the names vary even in the same township.

Children and the elderly are at the highest risk for As exposure, or at least the most vulnerable to contamination in general. The identification of this vulnerable cohort was necessary in order to plan the socio-educational actions. Only 25.1% reported having children at home, and only 18.4% of respondents reported sharing a home with people over the age of sixty. As might be expected, in some less urbanized communities, most children play in the backyards and gardens, in direct contact with the soil (Table 9.5).

When responding to questions about their leisure activities and free time, the respondents also mentioned spending time in their yard, mainly in Santa Bárbara districts. The respondents were asked to mention sites where they had contact with dust. Only 27 (7.9%) reported not having contact with dust. Some points deserve to be highlighted. For example, in the town Santa Bárbara, only 1.7% of the respondents reported not having contact with dust, while in Nova Lima it was 6.1%. The more rural character of Santa Bárbara seem to expose the population to dust more than those of Nova Lima. In Mina d'água, 5% mention their dust contact in respect to mining activities.

Table 9.4. Sanitation sewage services.

Sewage service options														
		Sewage network		Septic tank		Open skies		Rivers		Unknown		Tailings dam		Total
District	Township	n	%	n	%	n	%	n	%	n	%	n	%	n
Santa Bárbara	Barra Feliz	10	11.1	7	7.8	4	4.4	69	76.7	–	–	–	–	90
	Brumal	7	9.9	10	14.1	1	1.4	52	73.2	1	1.4	–	–	71
	Santa Bárbara city	3	60.0	1	20.0	–	–	1	20.0	–	–	–	–	5
	Total	**20**	**12.0**	**18**	**10.8**	**5**	**3.0**	**122**	**73.5**	**1**	**0.6**	–	–	**166**
Nova Lima	Nova Lima city	13	81.3	1	6.3	–	–	2	12.5	–	–	–	–	16
	Galo Novo/Velho	13	30.2	4	9.3	–	–	26	60.5	–	–	–	–	43
	Mingu de Cima/ de Baixo	29	61.7	5	10.6	3	6.4	10	21.3	–	–	–	–	47
	Matadouro/Isolamento	18	47.4	2	5.3	1	2.6	15	39.5	1	2.6	1	2.6	38
	Mina d'Água	18	54.5	1	3.0	1	3.0	6	18.2	–	–	7	21.2	33
	Total	**91**	**51.4**	**13**	**7.3**	**5**	**2.8**	**59**	**33.3**	**1**	**0.6**	**8**	**4.5**	**177**

Table 9.5.   Daily children's play and leisure sites (in percentages).

District	Township	In-home	Backyard	Land	Rivers	Club	Town
Santa Bárbara	Barra Feliz	83.1	71.2	57.6	22.0	22.0	15.3
	Brumal	77.5	82.55	35.0	30.0	22.5	27.5
	Santa Bárbara city	75.0	100.05	50.0	–	–	25.0
	Total	**80.6**	**76.7**	**76.7**	**24.3**	**21.4**	**20.4**
Nova Lima	Nova Lima city	85.7	28.6	–	–	14.3	57.1
	Galo Novo/Velho	46.7	86.7	60.0	26.7	20.0	13.3
	Mingu de Cima/de Baixo	73.7	52.6	5.3	5.3	15.8	47.4
	Matadouro/Isolamento	38.9	38.9	22.2	16.7	55.6	61.1
	Mina d'Água	54.5	50.0	31.8	9.1	9.1	77.3
	Total	**56.8**	**53.1**	**25.9**	**12.3**	**23.5**	**53.1**

c) Food habits. When questioning the participants about food habits, specifically regarding food origin, it was found that the residential gardens are part of everyday life in the communities under study. About 40% of respondents reported this type of cultivation, mostly irrigated by water drawn from springs. As mentioned previously, the water usually does not receive any kind of treatment, which may increase human exposure. Less than 30% eat fish once a month and the majority (53.1%) do so only sporadically. Among those who eat fish, about 30% fish in rivers and lakes from the region.

d) Respondents' occupational profile. Of those who completed the questionnaire, 36.2% work outside home, 30.3% work at home, and 4.7% are retired or have no occupation due to health problems. Although an association cannot be shown, it should be noted that some people retire with symptoms that might be due to As contamination: high blood pressure, stroke, cardiac disease, eye and lung problems, and anemia (▶ 2).

Of the 90 respondents (47 in Santa Bárbara and 43 in Nova Lima), 26.2% reported working in mining. In Santa Bárbara, of those who work or have worked in the mining business 49% mentioned the São Bento Company. In Nova Lima, 81.6% mentioned the Morro Velho Mining Company/Anglogold Ashanti. The major percentage among the respondents who work or have worked in mining in their regions, live in the townships of Galo Novo/Galo Velho (34.9%), followed by Barra Feliz (32.1%). Of this group, 32.6% reported not having passed any regular health assessment. Most (87.2%) reported being members of the Commission for Prevention of Accidents (Comissão Interna de Prevenção de Acidentes, CIPA). The percentage of those who work or have worked in the metallurgic industry is lower (9.1%). In this sector, 35.5% of the respondents stated not having passed any regular health assessment, and 80.6% reported participating in CIPA.

In spite of the relatively low percentage of respondents directly involved with the two main economic activities, 79.3% reported having parents or other relatives who had worked, or were working, in mining or metallurgy. The research also confirmed that these two types of work were the main income sources for 82.8% of the respondents' families. The results do not differ when analyzed by townships. In most townships, at least 80% of relatives work, or have already worked in either of these two activities. These data are relevant, as they indicate the influence of these two economic activities in these communities. The economic relevance possibly affects the perception of residents of the impact of mining and metallurgy – positive or negative.

9.3.2   *"Trade-off" measures*

To check the respondents' level of commitment to environmental protection measures in their communities, three questions were posed. The first question asked whether the companies should "stop their activities in the region even though this might generate unemployment". The majority (65.6%) disagreed. Still, 28.9% agreed in theory with this radical option. The answer shows the importance of economic influence for the survival of these communities that want environmental protection, but not at any cost, especially if such protection interferes with social and economic issues that affect people's lives.

When confronted with a radically contrary position, where it is stated that "*the companies must stay in the region, as there is not a direct relation between As contamination and damage caused to human health*", the respondents were divided. Consistent with the previous statement, 29.4% disagreed with this "liberality", and only 31.2% categorically agreed with the hypothesis. Therefore, the percentage of those who choose not to take a position increased from 5.5% regarding the first statement, to 39.4% for the second. A complementary trend emerged to the first response: the wish to satisfy demands and social issues affecting the community without relieving the companies of their environmental responsibility. This position on the second statement is reinforced by the answers to the third alternative, which argued that the companies "*must remain in the region, informing the population about the risks of As contamination and taking measures to avoid it*". In this case the correlation rises to 95%. This picture is repeated in townships and districts.

9.3.3 *Health*

Of the respondents, 19.8% reported having had miscarriages in their families, and 19.2% in their communities. Incidence was higher in Santa Bárbara with 24.2% and 20.5%, respectively, than in Nova Lima with 16.9% and 18.1% respectively (Table 9.6). In Santa Bárbara, 60% of the respondents in the city reported miscarriages in their own families (small number of respondents). In the other districts, however, the numbers were more significant: in Barra Feliz, 24.4% of respondents reported miscarriages in their families and 22.2% in their communities. In Nova Lima, the district Mina d'Água was negatively highlighted by the percentage of events in the community (48.5%), while in Matadouro/Isolamento, 23.7% of the responses indicate occurrences among relatives. While these data have to be interpreted with caution, they do suggest a high incidence of spontaneous abortions, whose accuracy and causes are beyond the scope of this study. Independent of the interpretation given to this information, it is noteworthy that none of the districts had a miscarriage rate below 10%, which deserves further study.

Regarding cancer cases, approximately 32.4% reported the occurrence in their families (Table 9.7; ▶ 2). The numbers indicate that the higher occurrences include stomach, uterine and breast cancer. The data shows that the percentage of lung cancer in Nova Lima

Table 9.6. Occurrence of miscarriages in the family and in the community.

District	Township	In family				In community			
		No		Yes		No		Yes	
		n	%	n	%	n	%	n	%
Santa Bárbara	Barra Feliz	68	75.6	22	24.4	70	77.8	20	22.2
	Brumal	58	81.7	13	18.3	58	81.7	13	18.3
	Santa Bárbara city	2	40.0	3	60.0	4	80.0	1	20.0
Nova Lima	Galo Novo/Velho	37	86.0	6	14.0	38	88.4	5	11.6
	Mingu de Cima/ de Baixo	40	85.1	7	14.9	39	83.0	8	17.0
	Matadouro	29	76.3	9	23.7	35	92.1	3	7.9
	Mina d'Água	28	84.8	5	15.2	17	51.5	16	48.5
	Nova Lima city	13	81.3	3	18.8	16	100.0	–	–

(10.7%) is much higher than in Santa Bárbara (1.8%). For skin and head/brain diseases, the percentages are the same in the two districts, 4.5% and 2.7% respectively. In Santa Bárbara, cancer of esophagus and throat are nearly double of those of Nova Lima, 8.9% and 4.5%. Table 9.8 shows these results by district.

Regarding the health services coverage, the public network assisted the vast majority of respondents: 75.2% by the Unified Health/Sistema Único de Saúde (SUS), and 2% by the Institute for Public Security of the State of Minas Gerais/Instituto de Previdência dos Servidores do Estado de Minas Gerais (IPSEMG).

These data about medical care imply that all educational campaigns should consider the involvement of the public health service, either with the support in monitoring symptoms found in the individuals, or with contributions to information and education.

9.3.4   *Perception of As related problems and level of knowledge and information*

The fact that the communities are very dependent on the two main economic activities in the region – mining and metallurgy – would certainly influence the degree of residents' awareness of arsenic, since these activities are more directly involved with the issue. In general, 77.8% of the respondents stated having some knowledge about the As issue, with no significant gender differences (80.6% men and 79.3% women). This picture changes when the length of time that the respondent has lived in the region and the respondent's age is taken into account. Knowledge of the subject increases with the respondent's age and the time living in the communities (Table 9.9).

Table 9.7.   Cancer occurrence in the families.

	Frequency	Percentage	Valid percentage	Accumulated percentage
No	232	67.6	67.6	67.6
Yes	111	32.4	32.4	100.0
Total	343	100.0	100.0	–

Table 9.8.   Types of cancer identified by township.

District	Township		st	br	ut	th	un	sk	le
Santa Bárbara	Barra Feliz	N	8	3	8	7	8	2	–
		%	13.3	5.0	13.3	11.7	13.3	3.3	–
	Brumal	N	11	11	2	3	2	3	5
		%	18.3	18.3	3.3	5.0	3.3	5.0	8.3
Nova Lima	City	N	4	2	–	–	1	–	–
		%	7.7	3.8	–	–	1.9	–	–
	Galo Novo/Velho	N	5	–	2	–	6	–	1
		%	9.6	–	3.8	–	11.5	–	51.9
	Mingu de Cima/ de Baixo	N	7	4	6	3	1	2	–
		%	13.5	7.7	11.5	5.8	1.9	3.8	–
	Matadouro	N	5	–	2	–	3	–	–
		%	9.6	–	3.8	–	5.8	–	–
	Mina d'Água	N	4	–	5	2	2	3	–
		%	7.7	–	9.6	3.8	3.8	5.8	–

st: stomach; br: breast; ut: uterine; th: throat; un: unknown; sk: skin; le: leucaemia

These data are consistent with the respondents' experience and their knowledge about the history of the community. In contrast to this apparent dissemination of knowledge about a problem that may affect them directly or indirectly, however, there seems to be little application of that information for daily behaviour. Among those who declared some awareness, only 11.6% considered themselves to be "well informed" or "very well informed". The majority (57.8%) considered themselves to be "ill informed", or they did not know, or had never heard about the issue.

The data indicate that the knowledge about arsenic is insufficient for people who have to live with a potential risk. In respect to the level of information, a small difference between men and women was detected. Men considered themselves to be better informed (15.7% versus 9.7%; Table 9.10).

Those that reported to have some knowledge were asked about possible effects of arsenic on human health. The questions allowed multiple answers. The first answer investigated the real number of respondents that were capable of identifying at least one of the possible As effects on the human health. Among the 267 respondents, who declared having some related knowledge, 45.2% (155) could not identify any problem. Skin and lung cancer (22.8%), and respiratory problems (19.9%) were highlighted among those who could mentioned some effect.

Another important aspect is how As-related information reaches the population. Friends and relatives were mentioned by 44.4% as their information source, as well as community meetings. Only 18.2% mentioned the companies (considering the answers "at work" and "technicians in the mining companies", the latter with 5.9%). This is somewhat disquieting since these communities depend on these companies.

Only one respondent mentioned the unions, CIPA and the "hospitals and healthcare system" as an information source. Schools appear in 9.6% of the answers, while "public system technicians" in 5.8%. Newspapers, magazines and television are also mentioned. However, these data show that, although some information reaches these communities, this does not happen in an organized and sequential way. The information is fragmented and dispersed, and often truncated.

Table 9.9.   Knowledge about arsenic, relative to respondent age (group).

Age		No	Yes	Total
17	n	6	8	14
	percentage	2.0%	2.6%	4.6%
24	n	12	28	40
	percentage	4.0%	9.2%	13.2%
35	n	15	54	69
	percentage	5.0%	17.8%	22.8%
49	n	22	68	90
	percentage	7.3%	22.4%	29.7%
89	n	6	84	90
	percentage	2.0%	27.7%	29.7%

Table 9.10.   Level of information, according to gender (in percentages).

Gender	Well informed	More or less informed	Ill informed
Male	15.0	42.6	41.7
Female	9.7	43.3	47.0

9.3.5   Perception of the key environmental problems

To understand the level of knowledge and the involvement of a population with its own life context, one can analyze the perception of the key problems in the community. Respondents mentioned those that usually are common to Brazilian towns and cities: water pollution (42%), air pollution (35.6%), and sewage (25.1%), which are covered in most the studies of environmental perception. It was to be expected that these issues received more attention from the respondents. Floods (22.4%), fish mortality (21%), and waste deposits (20.4%) were named as particularly relevant problems in the communities under study, albeit in small numbers.

The economic development mining and industrial, in both regions under study, certainly provoked many alterations, not only of the landscape but also in the management of the territory and its assets. Alteration of the water courses, in discharge and in quality occured. Water shortages were reported by 11.7% of the respondents, and those 21% who reported on fish mortality show a perception on the degradation level of the rivers in the project region. These environmental problems were also analyzed by township, to see possible differences between the two regions.

In Santa Bárbara, water pollution was perceived as a problem by 66.3% of the respondents, followed by fish mortality with 41%, and air pollution with 33.7%. Another negative highlight in Santa Bárbara relates to 13 respondents who mentioned problems with growing gardens and orchards, 11 from this town and only two from Nova Lima. In Nova Lima, fewer people viewed water pollution as a problem ("only" 19.2%). The most frequently mentioned issues in this town were air pollution (37%), tailing deposits (31.1%), and floods (26%). Comparing the data from the several townships, the problem ranking sees significant alterations.

One of the key problems in Santa Bárbara is obviously related to water (pollution). This does not emerge in Nova Lima in the same frequency. Fish mortality also appears strongly in the Santa Bárbara data, especially in Barra Feliz (44.4%). Here, the largest number of respondents (8) mentioned problems with vegetable gardens and orchards. The key problem in Nova Lima is seen in townships with tailings deposits (75.8%) and air pollution (66.7%) in the township Mina d'Água, besides floods in Matadouro/Isolamento (65.8%) and in Nova Lima city (43.8%).

9.3.6   Public participation and interest

The great majority of respondents expressed interest in receiving more information about environmental issues, particularly about arsenic, and only 6.7% were not interested. This is a positive backdrop for environmental education, since it shows that the population is concerned with activities that occur in their communities. In addition, most respondents showed a high level of participation and involvement in the actions developed by the community, albeit with little reference to any outside institutions.

When asked about institutions that have developed actions concerning the environment, a significant number of respondents reported not knowing any institution or developed action. Despite the respondents manifest interest, the success of any environmental education work will largely depend on a greater involvement of the institutions present in these communities, and ensuring continuity of the actions developed over time.

9.4   CONCLUSIONS

This study achieved its target by providing the institutions involved with important elements for intervention in those communities. The more detailed knowledge about the social reality, the residents' habits towards health and environmental aspects of the communities, is an essential instrument for more effective action of the public agents. At the same time it helps identifying the key issues for environmental education workshops (▶ 15).

From the analysis of the information collected, three aspects can be identified on perception, experience and knowledge of the communities:

- Contextual and social dimensions, which present multiple and diverse contexts
- Cognitive dimension of this reality through which the structure and forms of interaction with these contexts are understood
- Behaviour dimension, expression of the individual's interaction with their medium.

The participating communities were under similar conditions as the majority of other Brazilian municipalities. There is poor or non-existent access to services that often determines the level and quality of the local development, and has a direct and negative influence on the life and environmental conditions in many localities. The issue of the solid urban waste is an example, as well as sanitation and potable drinking water supplies. In the studied townships, the generally favorable economic conditions have not provided direct and universal benefits, such as full access to basic public services, even though these municipalities are part of the Iron Quadrangle, and therefore enjoy a strong economy with mining and processing-based metallurgy that certainly generates wealth for the region and the State.

On the macro-social dimension, the experience in the community is one of the most important factors in determining the level of knowledge about issues faced in daily life. Most respondents have lived for over two decades in the studied region, which explains their involvement with the community and related issues. However, this familiarity did not translate into active commitment mechanisms to better adapt to their exposure to environmental degradation and pollution.

The macro-social context does not unequivocally determine the perception of the people. People have inaccurate perceptions of reality, due to lack of information. For example, there can be serious implications when people, believe they are drinking potable water, when in fact there is potential for exposure to water-borne diseases. Changing these beliefs and perceptions is the only solution in some social contexts. With simple educational measures, often emphasizing personal hygiene, the health of the population can be ameliorated. By failing to properly identify a problem, as in the case of mistaken perception of water potability, the population may react by adopting behaviours that lead to increased risk of disease. This brings us to the behavioural dimension.

The communities usually resorted to alternative sources of water supply, with all negative implications to human health. This shows the importance of the awareness work. Failing to perceive the lack of treated water correctly, many families in the studied communities consumed water without any kind of treatment. Yet, access to treated water is one of the key indicators of family care and health. Since people also used untreated water when given a healthy alternative, socio-educational actions were proposed to minimize possible detrimental effects, and to reduce the risks to human health in these communities. The study clearly showed deficits in residents' behaviour, such as waste burning and non-treated water consumption, which could be addressed by environmental and health educational workshops. Similarly, personal hygiene and food habits need improvement (▶ 15).

The studied communities are mostly dependent on the main economic activity in the region, which certainly influences their perceptions and beliefs. Thus, the economic issue is likely to determine perceptions of the respondents and residents in relation to the positive and negative impacts generated in their communities. Another significant issue is the methods by which the respondents received information about the As issue. It should be noted that information was unavailable from the public sector, a situation that the ARSENEX project intended to address. Although some information reached these communities, this did happen in an organized and rational manner, but was fragmented and dispersed.

As stated above, problems such as water pollution, floods, fish mortality, air pollution, tailings deposits, and last, but not least, the problems with growing gardens and orchards, are strongly associated with the main economic activities in the studied regions. Therefore, the release of sewage directly into rivers can be attributed to the lack of investment in the improvement of public health services to these communities. The subject As exposure was infrequently mentioned in the answers to the questionnaire. Raising the profile of this issue is undoubtedly one of the main contributions of the ARSENEX project. Bringing it into the sphere of scientific discussions, and also of the res publica – the public affair – will arouse interest in participating in the debate.

Finally, it must be emphasized that, despite the interest in the completed research, the success of any environmental education work will depend upon the support of the institutions in these regions, to guarantee continuity. Prompt action is not enough to reach the proposed aims. Any joint and articulated action of these institutions needs not only to include the citizens, but also the private companies.

REFERENCES

FEAM: *Nosso rio, nossa gente: percepção e comportamento ambiental da população da bacia do Rio das Velhas – principais descobertas*. Fundação Estadual do Meio Ambiente. Belo Horizonte, 1998, p.80.

Jacobi, P.: A percepção dos problemas ambientais em São Paulo. In: Ferreira, L.C. & Viola, E. (eds): *Incertezas de sustentabilidade e globalização*. Editora Unicamp, Campinas, 1996, pp.177–188; ISBN 85-268-0387-5.

Norris, P. & Field, W.: Environmental public opinion: In: Hastings, E.H. & Hastings, P.K. (eds): *Index to international public opinion – 1979–1991*. Survey Research Consultants International; Greenwood Press, New York, 1992.

Section III
From air, water and soil to the human body

CHAPTER 10

Dust sampling and interpretation

Jörg Matschullat

The first chapter (▶ 1) presented an overview of global atmospheric As translocation and concentrations. While this is an important perspective, local and regional situations can be radically different from such global averages and demand much more specific attention. In 1998, a first attempt was made to obtain related data from the study area. There was no information available at that time, except for meteorological data provided by the Nova Lima station (▶ 7.3).

Atmospheric As emissions and transport occur in the gaseous and particulate form. Both organic and inorganic As species have been found in the gas phase (Hirner and Emons 2004; Mattusch and Wennrich 2005; Planer-Friedrich *et al*. 2006; ▶ 1). The most common and natural transport medium for As particles is dust. Most of the material normally consists of As-laden very small mineral particles (<<63 µm), such as pyrite (▶ 7.2, 12). In higher concentrations, and with decreasing diameter, these aerosols may represent a direct threat to plants, animals and humans, as the defense mechanisms of all organisms that directly inhale air are not resistant enough to fight the intake effects. Other forms of intake are hand-mouth contact (especially by young children under 6 years of age) or indirect ingestion of As-contaminated food (surface dust on vegetables and fruit). This ingestion may significantly contribute to the total human As exposure and is a subject of environmental and health education (▶ 2, 9, 14, 15).

10.1 SAMPLING (TOTAL, WET, DRY AND INTERCEPTION DEPOSITION)

To rapidly evaluate long-term exposure and at reasonable cost, active samplers (current situation) and passive samplers, including biomonitors (evaluation also of total retrospective load) can be used. Active sampling is more complex, especially if As species are to be preserved and when concentrations are relatively low (Rasmussen *et al*. 2007). The selection of the best-suited method depends on the sampling purpose and is subject to budget limitations and time availability. To obtain representative samples, their number per unit of space needs attention, since spatial heterogeneity may produce deviations by a factor 4 in a very limited area (e.g., 1,000 m²). Caution is needed to prevent non-authorized people from influencing samplers during operation or storage. The large deposition variability demands extended monitoring periods – a one year cycle can be seen as a minimum for outdoor measurements. Since inter-annual variability may be quite large, related studies should be performed. The selection of the sampler materials depends solely on the compounds to be determined. The following description refers primarily to inorganic compounds (trace metals).

The collection of **total deposition** is the simplest (and cheapest – between 10 and 20 € per unit) quantification of atmospheric deposition (Figure 10.1). Using this method, open samplers accumulate everything that deposits in a dry or wet form over a certain period of time. In theory, any open container will do, for example, a wide-mouth container with a capacity of 5 liters (HDPE – high density polyethylene) and a funnel at the top (Figure 10.1, left). A simple and inexpensive option has been successfully employed on the Kola peninsula, Russia (Reimann *et al*. 1997). The container is placed on a stable platform 1 m above ground, to avoid contact with suspended particles, especially during heavy rainfall and wind storms. It is

Figure 10.1.   Left: total deposition samplers in the field (front); right: Bergerhoff sampler (photo Kirsten Plessow).

even better to place the container on a PVC tube firmly planted in the soil. This will slow (bio)chemical changes of the sample. However, the biggest problem remains that chemical imbalances in the receptacle are uncontrollable, which can compromise the interpretation of the data. Still, for basic monitoring, this technique is widely used and trusted. The sampling intervals shall not exceed one week under tropical and subtropical conditions. The same is true for the Bergerhoff-style samplers. Using this method, glass beakers are used that allow for subsequent analysis of organic compounds. A metal protector prevents birds from defecating into the samplers (Figure 10.1, right).

To sample **wet deposition** only (e.g., rain), and to exclude the influence of dry deposition, wet-only collectors have to be used. At first glance, these look like rain gauges, but are equipped with a mechanism that automatically samples during precipitation events only (optoelectronic sensors). The materials used and the construction details depend on the given task. Commercial models are available (e.g., Eigenbrodt, Germany), but a fine workshop could build a customized model (Figure 10.2). Electrical energy is needed and regular maintenance is a requirement. The sampling intervals may not exceed a period of 2 to 4 weeks under standard conditions. Again, under tropical and subtropical conditions, intervals must be limited to one week to prevent deterioration and undesirable chemical reactions (and thus, sample alteration) inside the receptacle. Costs depend on the level of sophistication of the collector and may well exceed 10,000 € per unit. Custom-made models may start at 1,200 €, but require the related expertise of a workshop crew.

Dry deposition sampling can be rather expensive and complex (Rasmussen *et al.* 2007). Depending on the task, sampling intervals and duration, the first decision must be to select between passive or active samplers (low, LVS, and high volume samplers, HVS, for particulate material – PM 1, PM 2.5 and PM 10). Both types demand electric energy. The first type, preferably for outdoor use, operates with volumes around 1–3 m^3 of air per hour, while the other works with around 150 m^3 per hour (preferably indoor use, work place). Units of stacked filters allow for differentiating particle sizes if needed, and for the determination of radon. There is a large variety of commercial models on the market. Again, the best option might be a customized model, especially those developed by academic researchers (e.g., Willy Maenhaut's Gent sampler). A good workshop can build one for around 1,500 €, while commercial equipment easily costs above 5,000 €; Figure 10.3). Most samplers direct the air through inert filters. The filter residue can be analyzed after chemical digestion (ICP-OES, ICP-MS), after radiation (INAA, instrumental neutron activation analysis), or directly by solid-sampling (SS-GF-AAS – solid sampling-graphite furnace-atomic absorption spectrometry or LA-ICP-MS – laser ablation ICP-MS). The chemical composition can be determined with high accuracy per unit of time, as long as the sampler reliably counts the volume of air during sampling. Respective intervals are set for periods from hours (closed environment, HVS) to around a week (open air, LVS, Gent sampler, Partisol, etc.; Figure 10.3).

Passive samplers (e.g., Sigma-2; Kohler *et al.* 2007) are an alternative and a less expensive solution. These devices collect particles <10 μm on glass plates or sticky plates on the bottom of a settling tube with lateral openings, fitted into another tube, also with lateral openings. The ambient air is slowed down, and the particles are deposited in the centre of the sampler (Brown's molecular movement). The sampling plates should be replaced carefully, and the sample desorbed in the laboratory with diluted acids. Prior to sample desorption, the plate can be analyzed microscopically, which may reveal important information about sources of the material (biogenic, such as pollen; anthropogenic, e.g., originating from the high temperature combustion, or soot from diesel engines, etc.). The samplers are relatively inexpensive (250 € per unit) and have been successfully used from arctic (Greenland) to tropical environments (Figure 10.4).

Another type relates to **interception deposition**. This is the atmospheric input with mist and fog, increased by plants, especially trees and their large size canopies as interceptors (leaf

Figure 10.2.   Two models of wet-only samplers. Left: Eigenbrodt. Right: IÖZ model, workshop made (photo Jörg Matschullat).

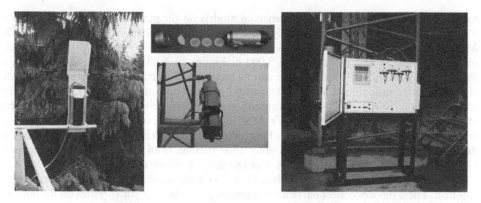

Figure 10.3.   Aerosol samplers. Left: Gent sampler, open. Right: Partisol© sampler (Rupprecht & Patashnick, USA) with individual components (upper inset) of the sampling cartridges (complete, lower inset): diffusion tubes, filters, honeycomb denuder (photo Kirsten Plessow).

Figure 10.4.   Sigma-2 sampler (on stake) in Mingu, Nova Lima (photo Jörg Matschullat).

area index). Interception may lead to a substantial deposition increase in the system (factor 100 to 1,000, related to the concentrations). Interception deposition may cause significant stress in plants and other biota. The simplest sampling method consists of nylon meshes or thin strings mounted on chemically inert frames and exposed to the air. It is necessary to prevent large particles from falling into the samples – and to prevent animals, like birds or monkeys, to sit or defecate onto the sampler. Sampler design includes a drain through which accumulated water passes into a container (usually a HDPE bottle). Such passive samplers are cheap and can be easily made (workshop). There are active and much more sophisticated and expensive fog samplers on the market. An artificial and very light air stream gently pushes humid air to a nylon harp but the collection has a lower degree of efficiency. However, these devices require electricity and the construction is more difficult and more expensive. Both types need very regular and careful maintenance.

Interception samples, and total deposition samplers, collected in inert plastic containers, should be transported to the laboratory and treated like any other water sample. It is recommended that air temperature, electrical conductivity and the pH values of an aliquot be measured at the sampling site. With small sample volumes – such as mist – only miniaturized sensors and electrodes should be used. Last, but not least, biomonitoring methods must be taken into account (see below). While their results may not necessarily be as precise, they certainly deliver significant information about atmospheric deposition. The ARSENEX project included biomonitoring with lichens, low volume aerosol samplers and domestic dust collection.

10.2 BIOMONITORING WITH LICHENS

At the beginning of the ARSENEX project, a simple and reliable method was sought to very quickly obtain data on As deposition and distribution in the sub-target areas around Nova Lima and Brumal (Santa Bárbara district). Lichens (lat. *lichenes*) are symbiotic organisms with a fungal and an algal partner. Related species occur all around the world under highly variable environmental conditions and with a great variety of generae, species and sub-species. Lichens grow slow and have neither roots nor rootlets; their metabolism is adapted to make efficient use of air-borne materials, including humidity. Among the different functional types of lichens, epilithic ones were selected. Unlike epiphytic and others, epilithic lichens grow on rock and stone surfaces, including man-made structures such as walls and roofs (Figure 10.5). The advantages of epilithic over epiphytic lichen (those growing on plants surfaces) is that they do not take up plant material or, e.g., water that has been in exchange with plant surface material. The lichen can be collected from upright stones or rocky surfaces, taking care to prevent standing water from affecting the specimen. Under favorable environmental conditions,

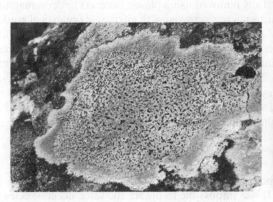

Figure 10.5. Left: close-up of an epilithic crustose lichen; right: lichen on a horizontal surface, Old English cemetary, Nova Lima (photo Jörg Matschullat).

lichens have a long lifetime (decades to centuries). Therefore, it takes no more than a sample to evaluate the air quality for extended retrospective periods of time. The lichen samples were collected in April 1998.

Nutrients will be absorbed by the organism and contribute to its metabolism. Therefore, nutrients cannot be used in evaluating atmospheric deposition. However, non-essential elements such as arsenic (and many other trace elements) can ideally be monitored. These trace components can be stored in individual cells and are protected from metabolism; an indirect detoxification mechanism. This type of storage appears rather stable, and the elements remain in the lichen body (thallus) until it perishes (Goyal and Seaward 1981; Matschullat *et al.* 1999; Puckett 1988; Richardson 1992).

Therefore, lichen can serve as biomonitors to evaluate atmospheric deposition in retrospect. The determined concentrations of non-essential compounds can be directly (but not linearly) related to the integrated deposition during the lifetime of the organism. When investigating spatial and temporal characteristics, however, the age of an organism (besides its location) must be known and individuals of similar dimensions should be sampled (to obtain material of similar age). In respect to temporal resolution, this continues to be unsatisfactory, as the lichen growth is directly related to the microclimatologic situation and to local nutrient supply. Therefore, results obtained from lichen that grow freely must be interpreted in a less quantitative way. Other aspects require attention: the lichen species digestion efficiency (dissolution in the laboratory) and individual behavior (growth rates etc., metabolism) are very specific. To compare atmospheric deposition appropriately on the local and regional levels, only specimen from the same species may be sampled (Matschullat *et al.* 1999). The difficulties illustrate that results from this method must be interpreted as indicative only.

10.2.1   *Sampling, sample preparation and analysis*

Epilithic lichens (*Xanthoparmelia farinose* and *Parmotrema grayanum*) were collected in both the Nova Lima and Santa Bárbara districts from vertical walls and tombstones. The advantage of tombstones lies in the fact that lichens can be easily sampled from the smooth stone surfaces and that the burial age, and thus the maximum age of the lichen is known. The thalli were carefully scraped from the rock surface with a high quality stainless steel blade (As free), so as not to extract material from the stone substrate. At some sites, where the specimens were very small, more samples were taken with various sub-samples of the same species, but on a larger area. The thalli parts were placed into small PE-Whirlpack® bags, labeled and stored in a dark place. The specimens were dried within 24 hours exposed to the air under thin laboratory paper sheets to protect them from contamination and deterioration. The material was ground (<63 µm) thereafter in a planetary mill (Pulverisette 5, Fritsch, Germany). Before grinding, mineral particles were carefully removed using plastic tweezers under a magnifying glass. Sample digestion was done with a microwave-oven with nitric acid (analytical grade and supra-pure), and a direct analysis of solids in parallel. After digestion, the liquid was diluted to 50 mL or 100 mL (depending on the weight of internal mass, 50 mg or 100 mg, respectively) and concentrations determined by GF-AAS (AAS5-EA, Analytik-Jena, Germany) and ICP-MS (Perkin Elmer Elan 5000). The direct analysis was done by SS-GF-AAS (AAS5-EA, Analytik-Jena, Germany). Quality control was carried out with standard reference material (IAEA 336, 0392) and sample replicates; the two methods were compared independently. Alternate samples were duplicated (two digestions) for better control.

10.2.2   *Results and interpretation*

A limited number (n = 11) of composite samples were collected in both towns, Nova Lima (n = 6) and Santa Bárbara (n = 5). It was impossible to obtain the same lichen species at all sites. This inhibits a direct comparison. The site in Santa Bárbara produced the same species, but not the same sub-species. Therefore, it is likely that the differences found in the two districts are real and the samples reveal a picture with substantial meaning. Background lichens tends to yield <1 mg As kg^{-1}. Average values were higher in both towns (Santa Bárbara 3.3 mg kg^{-1}, Nova Lima 17 mg kg^{-1}; Figure 10.6). This indicates that the degree of atmospheric As transport in both areas was above background levels. Since adjacent soils show a rather high As level (▶ 12), relatively high As concentrations are not surprising. The lowest concentrations were encountered around Brumal, directly related to the relatively lowest As values in the adjacent soils (▶ 12). The regional background can be taken from the nearby Caraça Nature Reserve (*Santuário do Caraça*) with minimum concentrations of 0.5 mg kg^{-1}. There, no elevated geogenic As occurs, while the nearby township Brumal represented elevated values (maximum 7 mg kg^{-1}). Similar variations were found in plants, herbs and vegetables (▶ 13), indicating the need to study positive anomalies with high spatial resolution. It is noteworthy, however, that species of *Parmotrema grayanum* and *P. reticularum* were sampled in the Caraça Reserve, species that were found only once in Brumal. Nevertheless, the encountered concentrations (also of other elements) confirm results from other remote regions, e.g., in Europe and Canada (Matschullat *et al.* 1999).

Average concentrations between 4 and 23 mg kg^{-1} were found in Nova Lima. The lowest values occurred in the northwest of the city (a recreation club at a site opposite to the prevailing winds that pass by the foundry and ore processing unit in the town centre of Nova Lima). These samples also showed lower concentrations of other trace-elements, such as Ag and Cu, than the other samples from Nova Lima. The highest concentrations were found in the city centre and in the downwind direction from the mining and foundry operations, the old municipal cemetery and in the English cemetery (Figure 10.6). The buildings at the As source were being demolished, a process that generated additional dust, typical of the current situation in the region. The modern mine constructions are located in a valley between Nova Lima and the neighbouring community Raposos, where no samples were taken.

The evaluation of air quality using epilithic lichen demonstrated its validity – and its limitations. In this study, conditions were not ideal, due to the different lichen species encountered between the sampling sites. Even so, the data are valid and useful. As anomalies emerge distinctly and occur only at sites with specific As sources (Matschullat *et al.* 1999). The results

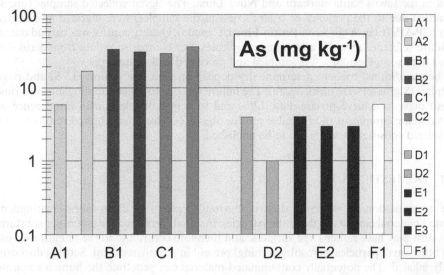

Figure 10.6.   Arsenic concentrations (mg kg^{-1}) in lichen from Nova Lima town (A-C) and Santa Bárbara district (D-F). A: Quintas Club; B: old municipal cemetery; C: old English cemetery; D: Caraça Nature Reserve; E: Santa Bárbara town; F: Brumal (Matschullat *et al.* 2007).

show that both districts receive an enhanced atmospheric As input, and that this flux seems to be considerably higher in Nova Lima. Since the biomonitoring using lichens does not allow for a more precise definition of whether the As load should be interpreted as a historic stress or if it is still active, other methods were needed to answer this question. Both direct particle sampling and the determination of total deposition could give an answer, but so could the analysis of soils and plants (▶ 12 and 13). As the adjacent soils maintain any As enrichment for long periods of time (decades to centuries), there is an almost permanent source that will redistribute arsenic not only through the water pathway, but through the lower atmosphere as well. It is short distance transport, not long-range transport, that explains the elevated level of As distribution variability in soils and plants. A dominance of long-range transport would lead to a homogenized As distribution in the local environment. The dominant As sources in both districts are gold mining and related tailings deposits that accumulated for centuries in the area and distribute arsenic in the local environment.

10.3   AEROSOLS

Both indoor and outdoor air may carry a substantial aerosol load. The size of suspended particles ranges from nanometers to about 100 micrometers. While large particles (>1 μm; dust) settle rather quickly (minutes to hours, depending on air mass movement and internal energy), fine particles (<<1 μm; PM1) can stay for extended time in the atmosphere (days) and are usually washed out of the atmosphere by rain.

To evaluate the results from the lichen samples, two collection campaigns were conducted to directly sample aerosols in ambient air. A low volume aerosol sampler (LVS) was mounted for several weeks on the COPASA pumping station near Morro do Galo during August 2000. Situated in a narrow valley, the location would allow capturing aerosol from the immediate neighbourhood. Potential sources identified there, included a large tailings deposit (Galo Novo) with very high As concentrations (▶ 12; now remediated), a large municipal open waste disposal site for Nova Lima (now deactivated), the ruins of the old As trioxide factory on Morro do Galo (now remediated) and the township Galo Novo with about 200 inhabitants.

Six passive aerosol samplers (Sigma-2) were set up in August 2003 at six locations for one week in the towns Santa Bárbara and Nova Lima. The eleven collected samples (one was discarded due to the presence of a bird's nest in the sampler) were digested and analyzed with ICP-MS (Elan 9000, Sciex-Perkin-Elmer, Canada). Quality control was carried out with standard reference material and sample replicates. For comparison, data from aerosol collected in Saxony, Germany, was sampled and processed in the same way.

These additional trials to determine dry deposition with low volume (LVS) and passive samplers (Sigma-2) were unsuccessful. The failure was due to short sampling intervals (insufficient mass to achieve precise data, LVS) and to very probable human interference with the sampling equipment (non-reliable results, Sigma-2). Thus, the achieved results were not considered consistent and are not to be published.

10.4   DOMESTIC DUST

The accumulation of domestic dust is inevitable (except in clean-room environments). Related material consists of domestic particles, materials from outside and other substances. This dust may have rather large surfaces, and therefore contribute to the retention of gases and other small particles (aerosol scavenging) present in the environment. Settling dust can be easily inhaled. The potentially contaminated material can penetrate the human respiratory system where toxins are easily absorbed (Rasmussen 2004).

To evaluate the potential exposure of people living near or on As-rich sites, 47 samples of domestic dust were collected in Nova Lima and Santa Bárbara. Using laboratory towels, dust on surfaces like window sills, furniture, etc., was collected from every room in the house, to obtain one composite sample per household (Figure 10.7). These samples were stored in sealed PE-Whirlpack® bags to await further processing. The towels were digested in concentrated nitric acid, and subsequently analyzed with ICP-OES (Optima 3000, Perkin Elmer, USA) and ICP-MS (Elan 9000, Sciex-Perkin-Elmer, Canada). The results from this sampling method were indicative only, due to a systematic error: the towels had not been weighed before use; thus the determined concentrations could be assessed as an estimate only (Raßbach 2005). To obtain representative results from the region, new samples were collected using a stainless steel blade. The samples were placed in HDPE sample holders and sealed for subsequent processing. The As concentrations were determined by Instrumental Neutron Activation Analysis (INAA; Corte 1986; Menezes *et al.* 2000). After sample inweight and conditioning in polyethylene tubes, the material was irradiated for 4 hours in the nuclear research reactor TRIGA MARK I IPR-R1 at the Centro de Desenvolvimento da Tecnologia Nuclear / Comissão Nacional de Energia Nuclear (CDTN / CNEN), with a mean thermal neutron flux of 66×10^{11} neutrons cm^{-2} s^{-1} at 100 kW. Gamma spectrometry was performed after decay of interfering radionuclides (Table 10.1).

Obviously, the As concentrations in home dust were not directly related to the As concentrations in the local soil. It must be considered, however, that the part of Matadouro that is located on the old tailings deposit was sealed with asphalt, while the buildings in Galo Novo and Galo Velho sat directly on the open soil back then. In the meantime, the local government has successfully remediated the Galo Novo tailings deposit (▶ 17).

The encountered As-enriched aerosols and dusts can easily be explained by the local As concentrations in the soils (▶ 12). The concentrations in Nova Lima were almost ten times higher than those in Santa Bárbara, reflecting the general difference between both districts in respect to As pollution (▶ 10–14). Physical movement on the soil surface, whether caused by wind, animals, people walking, or vehicles, redistributes the contaminated material. Human intake (inhalation and ingestion) can be increased by hand-mouth contact, particularly with younger children (▶ 14).

Figure 10.7.   In-door dust collection with laboratory wipes (▶ 11). Photo Michelle Amorim.

Table 10.1.   Arsenic in domestic dust and local soil in Minas Gerais.

District	Township	As (mg kg^{-1})	As in local soil (mg kg^{-1})
Santa Bárbara	Santana do Morro/ Barra Feliz/Brumal	21.8–28.7	37–83
Nova Lima	Galo Novo	0.7–3.5	360
	Galo Velho	11.0	170
	Matadouro	1.5	1,600

REFERENCES

Borba, R.P., Figueiredo, B.R., Rawlins, B. & Matschullat, J.: Arsenic in water and sediment in the Iron Quadrangle, Minas Gerais, Brazil. *Revista Brasileira de Geociências* 30:3 (2000), pp.554–557

Corte, F. de.: *The k0-standardization method: a move to the optimization of neutron activation analysis.* Rijksuniversiteit Gent, Faculteit van de Wetenschappen, 1986, p.464.

Goyal, R. & Seaward, M.R.D.: Metal uptake in terricolous lichens. I. Metal localization within the thallus. *New Phytologist* 89:4 (1981), pp.631–645.

Hirner, A.V. & Emons, H.: *Organic metal and metalloid species in the environment – analysis, distribution, processes and toxicological evaluation.* Springer Verlag, Heidelberg, 2004, p.328.

Kohler, F., Mölter, L., Schultz, E., Dietze, V., Schütz, S. & Helm, H.: Passive sampler Sigma-2 as an inlet for an optical aerosol spectrometer. *European Aerosol Conf. Salzburg*, 9–14 September 2007.

Matschullat, J., Scharnweber, T., Garbe-Schönberg, D., Walther, A. & Wirth, V.: Epilithic lichens – atmospheric deposition monitors of trace elements and organohalogens? *J. Air Waste Manage. Assoc.* 49:10 (1999), pp.174–184.

Matschullat, J., Birmann, K., Borba, R.P., Ciminelli, V.S.T., Deschamps, E.M., Figueiredo, B.R., Gabrio, T., Haßler, S., Hilscher, A., Junghänel, I., de Oliveira, N., Schmidt, H., Schwenk, M., de Oliveira Vilhena, M.J. & Weidner, U.: Long-term environmental impact of As-dispersion in Minas Gerais, Brazil. In: Bhattacharya, P., Mukherjee, A.B., Bundschuh, J., Zevenhoven, R. & Loeppert, R.H. (eds): *Arsenic in soil and groundwater environments: biogeochemical interactions. Trace metals and other contaminants in the environment*, Volume 9. Elsevier, Amsterdam, 2007, pp.365–382.

Mattusch, J. & Wennrich, R. (eds): Novel analytical methodologies for the determination of arsenic and other metalloid species in solids, liquids and gases. *Microchim. Acta* 151:3–4 (2005), pp.137–275.

Menezes, M.A., de, B.C., Sabino, C., de, V.S., Amaral, A.M. & Pereira Maia, E.C.: k0-NAA applied to certified reference materials and hair samples: evaluation of exposure level in a galvanizing industry. *J. Radioanal. Nuclear. Chem.* 245:1 (2000), pp.173–178.

Planer-Friedrich, B., Lehr, C., Matschullat, J., Merkel, B., Nordstrom, D.K. & Sandstrom, M.W.: Speciation of volatile arsenic at geothermal features in Yellowstone National Park. *Geochim. Cosmochim. Acta* 70:10 (2006), pp.2480–2491.

Puckett, K.: Bryophytes and lichens as monitors of metal deposition. In: Nash III, T. & Wirth, V. (eds): *Bibliotheca Lichenologica: Lichens, Bryophytes and air quality*. E. Schweizerbart, Stuttgart, 1988, pp.231–267.

Rasmussen, P.E.: Elements and their compounds in indoor environments. In: Merian, E., Anke, M., Ihnat, M. & Stoeppler, M. (eds): *Elements and their compounds in the environment*, Volume 1:11. Wiley-VCH, Weinheim, 2004, pp.215–234.

Rasmussen, P.E., Wheeler, A.J., Hassan, N.M., Filiatreault, A. & Lanouette, M.: Monitoring personal, indoor, and outdoor exposures to metals in airborne particulate matter: Risk of contamination during sampling, handling and analysis. *Atmos. Environ.* 41:28 (2007), pp.5897–5907

Raßbach, K.: *Arsentransfer und Human-Biomonitoring in Minas Gerais, Brasilien.* Unpubl. M.Sc. thesis, TU Bergakademie Freiberg, 2005, p.93 + annex.

Reimann, C., de Caritat, P., Halleraker, J.H., Volden, T., Äyräs, M., Niskavaara, H., Chekushin, V.A. & Pavlov, V.A.: Rainwater composition in eight Arctic catchments in northern Europe (Finland, Norway and Russia). *Atmos. Environ.* 31:2 (1997), pp.159–170

Richardson, D.H.S.: Pollution monitoring with lichens. In: Corbet, S.A. & Disney, R.H.L. (eds): *Naturalists' handbook*, Volume 19. Richmond Publishing Co. Ltd., Slough, Great Britain, 1992, p.76.

CHAPTER 11

Surface water

Olívia Vasconcelos, Sandra Oberdá, Eleonora Deschamps & Jörg Matschullat

The hydrosphere is often seen as the most active medium for As transfer to man. Evidence derives from numerous incidents, mainly in the Americas and Asia, namely Argentina, Bangladesh, Chile, China, India, México, Mongolia, Taiwan, the USA, and West Bengal, from where cases of human intoxication through water have been reported (e.g., Bundschuh et al. 2008a, b; Chappell et al. 2003). From a toxicological perspective, the inorganic reduced $As^{(III)}$ is more toxic than the oxidized $As^{(V)}$ (▶ 2), making anoxic (ground)waters more risky than aerobic surface water. Arsenic occurs as arsenate in most surface and drinking waters, due to the presence of sufficient dissolved oxygen (▶ 1). In anaerobic waters, arsenides can occur (Goyer et al. 1999; Irgolic 1994). Arsenic mobilization and translocation to waters follow various paths:

a. Direct transport by surface and percolation water is the main factor for the transport of suspended and dissolved solids. This includes the As mobilization from primary ores and mining-related activities. Inadequate control of tailing dams, combined with high precipitation can cause significant losses of waste material into surface waters.
b. Percolation of rain water through waste deposits in unsealed areas leads to As lixiviation and results in groundwater contamination.
c. Heavy rains may transport arsenic in both dissolved and particulate forms directly from As-contaminated soils to the water courses.

As a result, the investigation of various types of water, from groundwater via all types of surface water, including mine-discharge and semi-legal abstraction of all sorts of untreated surface water by individual households had to be scrutinized. Water samples were collected in the Nova Lima and Santa Bárbara districts, with emphasis on the das Velhas (▶ 7) and the Conceição (tributary to Rio Piracicaba) rivers.

11.1 SAMPLING, SAMPLE PREPARATION AND ANALYSIS

Methods for collection and preservation of water samples varied depending on sample type. Water samples from rivers were collected with HDPE buckets attached to a plastified extension rod, or directly scooped with the bucket; always at a certain distance from the source (spring, well, etc.) or from river banks. The bucket was previously washed with diluted nitric acid and rinsed several times before each sampling.

Determinations of pH-value, temperature, electrical conductivity (μS cm^{-1}), dissolved oxygen (mg L^{-1}) and turbidity (NTU) were measured in situ with a calibrated multi-parameter probe (Horiba U-10, Japan).

To determine cations and anions, samples were filtered immediately in situ with polycarbonate (PC) membrane filters (0.45 μm, Schleicher and Schüll, Germany and Nalgene, USA) in a 300 mL PC filter holder (Sartorius, Germany), using a hand pump (Nalgene, USA). (Lotic water often shows high content of suspended solids. This may slow down filtration considerably, up to 30 minutes for 200 mL – and needs to be considered). After filtration, the liquid was divided in two aliquots, one remaining non-acidified (30 mL) for anion analysis; the other acidified (2 mL HNO_3 p.a. per 100 mL) for cation analysis. All samples were stored

119

in previously acid-washed PTFE and HDPE flasks, stored in the dark at a temperature <4°C and transported refrigerated to the laboratory.

The anions (F^-, Cl^-, NO_3^-, PO_4^{2-} and SO_4^{2-}) were determined by ion chromatography (DX-120; Dionex, USA), and the cations (Na^+, K^+, Ca^{2+}, Mg^{2+}, total Fe and total Zn) were quantified by ICP-OES (Perkin Elmer Optima 3000, USA). The determinations were done in duplicate, and quality control included the analysis of certified reference material (NIST 1643 D).

Arsenic and other trace elements were determined with GF-AAS (graphite furnace-atomic absorption spectrometry) and ICP-MS (Elan 9000, Perkin Elmer-Sciex, Canada), respectively, with detection limits considerably below the maximum limits established by the environmental law.

Water for the microbiological determinations (total coliforms, fecal coliforms and fecal streptococcus), was collected in autoclaved polyethylene flasks and the samples were stored at a temperature of 4°C for 8 hours prior to analysis.

11.2   RESULTS AND INTERPRETATION

Distinct differences between the two hydrographical basins become visible in the obtained data (Table 11.1) and motivated the differentiated display in two sections, one for the das Velhas river basin, the other for the Piracicaba basin (▶ 7). Due to the very large variability e.g., of water chemistry because of seasonal discharge variances, repeated sampling is a must. The presented data are averages (and ranges of averages over a six-year period, 1998 to 2003, in parts longer) involving mostly irregular sampling during both wet and dry seasons.

11.2.1   Das Velhas river basin

Most of the water supply of Belo Horizonte comes from the das Velhas river. A total of 31 surface water samples were collected along the river in the Nova Lima district in both dry and wet season (Table 11.1). In addition, more than 1,000 samples were collected for an independent monitoring program (Table 11.2). The water temperatures reflect the annual variability, with lower averages in the Santa Bárbara district (▶ 11.2.2). Electrical conductivity is low, but typical for the environment. While maximum dissolved oxygen levels show saturation and thus, very good conditions, individual samples revealed critically low values. Both biological (BOD) and chemical oxygen demand (COD) lead to oxygen sags, particularly at times of waste discharge and of higher temperatures.

Table 11.1   Water quality of das Velhas (NL) and Piracicaba rivers (SB), global background (Reimann and Caritat 1998) and maximum permissible limits for surface water (#COPAM 1986).

Parameter	Range NL	Range SB	Background	Limit values[#]
T (°C)	16–29	14.6–24.0	–	–
pH-value	6–7.8	4.0–7.5	–	6–9
el. cond. ($\mu S\ cm^{-1}$)	4–140	6–112	–	–
Diss. O_2 (mg L^{-1})	3.5–10.3	5.0–10.6	–	5.0
F^- (mg L^{-1})	<0.01–2.2	<0.01–1.35	0.001	1.4
Cl^- (mg L^{-1})	<0.1–11.9	<0.1–1.5	8	250
NO_3^- (mg L^{-1})	0.19–60	<0.1–43.5	–	10
SO_4^{2-} (mg L^{-1})	0.28–1,046	0.01–44	4	250
Ca^{2+} (mg L^{-1})	0.5–410	0.17–6	18	–
Mg^{2+} (mg L^{-1})	0.2–111	0.04–2.3	4.1	–
Na^+ (mg L^{-1})	<0.1–60	0.06–0.76	6.1	–
K^+ (mg L^{-1})	0.02–10.6	0.02–0.8	2.3	–
Total Fe (mg L^{-1})	<0.01–0.46	<0.01–0.23	0.04	0.3
Total Mn (mg L^{-1})	<0.05–3.2	<0.005–0.053	0.004	0.1
Colif. (NMP 100 mL^{-1})	285–2,400	1–24,000	–	1,000

Table 11.2   Simple result statistics of water samples, das Velhas river basin.

Parameter	# of samples	Median	Standard deviation
Upper course			
Total As (mg L^{-1})	506	0.020	0.049
el. conductivity (μS cm^{-1})	660	198.3	461.9
Dissolved solids (mg L^{-1})	632	175.2	436.7
Suspended solids (mg L^{-1})	667	135.9	392.3
Total solids (mg L^{-1})	667	304.9	571.4
Turbidity (mg L^{-1})	667	142.2	455.8
Temperature (°C)	667	22.0	2.9
Sulphate (mg L^{-1})	337	100.7	294.8
pH-value	667	7.0	0.4
Total river basin			
Total As (mg L^{-1})	1,004	0.022	0.044
el. conductivity (μS cm^{-1})	1,312	161.9	115.5
Dissolved solids (mg L^{-1})	1,197	108.2	67.8
Suspended solids (mg L^{-1})	1,338	154.1	396.2
Total solids (mg L^{-1})	1,342	261.3	402.9
Turbidity (mg L^{-1})	1,342	163.5	457.3
Temperature (°C)	1,342	23.7	3.2
Sulphate (mg L^{-1})	652	9.1	7.0
pH-value	1,342	7.2	0.5

Projeto Águas de Minas (2008).

Comparing the surface water results with the limits established by the COPAM norm, fluoride, nitrate, sulfate and total manganese sometimes exceeded these limits (Table 11.1). While elevated Mn values will mainly be related to the above-average Mn-concentrations in many rocks and soils of the area, the enriched anion values most likely reflect mining activities (SO_4^{2-}, possibly F$^-$) and human refuse (NO_3^{-1}). Quite obviously, the latter is corroborated by rather high values for coliform bacteria that pose a serious health risk to the population.

Pollution effects from the upper das Velhas river basin occur all along the river with bad water quality and frequent fish death. The highest toxin values were regularly found in samples from the Cardoso stream, also known as "Água Suja" (dirty water stream), and in the Queiroz creek (draining current mining and beneficiation operations), both das Velhas tributaries in the Nova Lima district. To date, the das Velhas river has mostly been able to dilute the contaminants, but this certainly is not a sustainable practice. As mentioned above, tailings deposits from the Morro Velho Mine occupy the sub-basin of the das Velhas river and lie along the Cardoso stream:

- **Fábrica de Balas** deposit, right bank, occupied by Anglo Gold (MMV) installations and buildings (average depth 7 m; area 14,800 m²; volume 121,923 m³; As level 27,430 mg kg^{-1}). Part of this deposit was removed 1994–1996 by Nova Lima City Hall to make room for a road. The creek was channelized and the entire region urbanized. The material was sent to the Isolamento deposit,

- **Madeira** deposit, left bank, opposite to the Fábrica de Balas deposit (same changes as above; average depth 6.6 m; area 25,800 m²; volume 130,765 m³; As level 29,495 mg kg^{-1}),
- **Rezende** deposit, left bank; partially removed during the works of Nova Lima City Hall and transferred to Isolamento (maximum depth is 11.34 m; area 10,625 m²; volume 37,633 m³; As level 22.670 mg kg^{-1}). Part of the area is occupied by a vocational school, a multi-sport gymnasium and roads,
- **Matadouro** deposit, right bank, small and long-time urbanized, a popular district called Areião do Matadouro (area 5,770 m²; volume 21,413 m³; average As level 25,117 mg kg^{-1}),

- **Isolamento** deposit, left bank (mean depth 7.32 m; area 32.175 m², volume 104,714 m³; average As level 22,526 mg kg⁻¹),
- **Galo** deposit, left bank, near the confluence with the das Velhas River. The last deposit along this river and the largest in the area (mean depth 14 m; area 58.900 m²; volume 619,950 m³; average As level 10,975 mg kg⁻¹).

In 1910, an arsenic trioxide (As_2O_3) factory was built on the right banks of Cardoso stream, opposite to the Galo deposit, and remained in operation until 1975. Arsenopyrite-rich ore from the Morro Velho Mine was used as raw material to produce As_2O_3 for wood preservation and seed protection. The chimney of the old factory contributed to As dispersion in the vicinity. Excavations revealed that the As level reached 24.6% in the deposited material around Morro do Galo. Groundwater contamination was inevitable, as As levels up to 41.4 mg L⁻¹ were detected in wells on the left bank of the das Velhas river. The surface water quality in the das Velhas basin was examined for arsenic with data from the water quality monitoring by the Minas Gerais Water Management Institute (IGAM) between 1997 and 2008. Individual critical total As values occurred along the das Velhas river and its tributaries. The highest average value was found in the Cardoso stream, near its confluence with the das Velhas river (BV062; Figure 11.1).

However, the Cardoso stream water, near its confluence with the das Velhas river (BV062), shows a decreasing As concentration trend (Figure 11.2). Arsenic sources in the basin are concentrated in the upper course of the das Velhas river, where the economy is based on iron, gold and gem stone production (▶ 6). To test variability and relationship of various water quality parameters (characteristic of diffuse and point pollution related to mining activities), the upper course of the river and its basin were evaluated separately to assess possible differences between the upper and lower reaches of the river (Table 11.2).

The difference between both reaches is obvious and mostly statistically significant, except for the relation between arsenic and pH-values (Vasconselos *et al.* 2007). The upper course yields higher electrical conductivity, already explainable with the considerably higher sulphate concentrations and higher amounts of dissolved solids. The highest As concentrations occur in the upper reaches in the dry season, while the rainy season shows elevated values for total arsenic in the lower reaches, due to remobilization of previously sedimented particles. The rainy season shows a significant correlation between arsenic and dissolved solids, confirming the impairment of water quality in the das Velhas river by diffuse pollution.

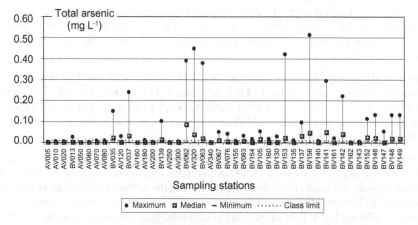

Figure 11.1.   Total arsenic in the das Velhas river basin from 1997 to 2008 (BV stations), and 2002 to 2008 (AV stations). Class Limit according to COPAM/CERH norm 10/2008 (IGAM 2008).

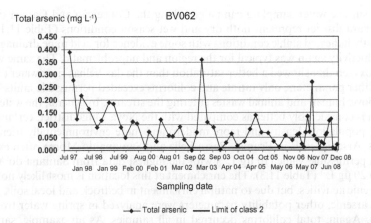

Figure 11.2. Total dissolved As (mg L⁻¹) in the Cardoso stream (BV062, see above) from 1997 to 2008. The horizontal line indicates the permissible level for surface water (10 µg L⁻¹).

The high As content in the Cardoso stream reflects both the natural As presence in the Nova Lima region, and the contribution by processing of Au-containing arsenopyrite. The highest values were related to sampling points close to the old Galo tailings deposit. At the beginning of the ARSENEX project, some of the Galo Velho residents lived in direct physical contact with the deposit, since part of its surface was used as a football (soccer) field and play area primarily for children. Other parts were used as vegetable gardens, cultivated in the backyards of the houses located on the surface of the deposit. Although most homes (except for one) were provided with treated water by COPASA (▶ 8), dwellers used water from springs and even mine discharge. The Project led to important mitigation actions that prevent further exposure (▶ 17). Today, the entire population of Nova Lima is supplied with quality-controlled potable water from COPASA. Therefore, untreated surface and ground-water are no longer used for domestic purposes. Related health risks, particularly in respect to coliform bacteria therefore should decrease considerably (▶ 14).

A few groundwater samples were taken by others (FEAM and British Geological Survey) without ever publishing the results due to limited representativity. Yet, those data give some insight and shall be partly reported here (Rawlins *et al.* 1997). Some related information based on the same report have been published by Williams (2001). Two samples each from both the Morro Velho environment (Nova Lima) and the Passagem de Mariana mine (Santa Bárbara, see below 11.2.2) were taken. The total As concentrations were between 11 and 1,700 µg L⁻¹, and As⁽ᴵᴵᴵ⁾ concentrations of 6 to 790 µg L⁻¹ at roughly at 50%. The pH-values were slightly above neutral and electrical conductivities between 190 and 250 µS cm⁻¹. Iron and Mn-concentrations were always below 0.1 µg L⁻¹, and 0.4 µg L⁻¹, mostly at least an order of magnitude below these maximum values. The elevated As concentrations in groundwater corroborate results from governmental monitoring in Minas Gerais, and can be seen as typical for anoxic groundwater in the gold mining areas.

11.2.2 *Piracicaba river basin*

The Piracicaba river, one of the Doce river tributaries, starts in the Espinhaço range with the Conceição and Caraça rivers as important tributaries in the study area. The Espinhaço range separates the Doce, São Francisco and Jequitinhonha river basins. Located in the central Doce basin, the Piracicaba basin is situated in the Rio Doce State Park (nature reserve), which is compromised by various economic activities of high environmental impact (iron ore mining and steel foundries, massive eucalyptus plantations for charcoal and cellulose production), and urban environment.

Twenty surface water sampling campaigns along the Conceição and Caraça rivers in the Santa Bárbara district represent both dry and wet season conditions (Table 11.1). The pH-values mostly indicated stable conditions with some evidence for acid mine drainage. The electrical conductivity again was typical for the region and unproblematic. The same was true for dissolved oxygen that showed a better saturation than the das Velhas river water (Table 11.1). From all other parameters, only nitrate and coliforms exceeded permissible limits – similar to reasons above, human and animal wastes entering the streams. In general, the water quality in this region is considerably better as compared with the Nova Lima district, certainly also due to a lower population density and more modern and more environmentally friendly mining operations. Arsenic levels in Brumal, Barra Feliz and Santana do Morro often exceeded the maximum permissible value (10 μg L^{-1}) with the highest data from Santana do Morro with more than 20 μg L^{-1} (Table 11.3). The enrichment at this location is most likely not caused by anthropogenic activities, but due to natural enrichment in bedrock and local soils.

Besides arsenic, other potability parameters were analyzed in spring water from Santana do Morro. Again, total coliforms occurred in all samples. As an example, samples from March 2007 yielded 2.2×10^3 NNP 100 mL^{-1} total coliform and 7.4 NMP 100 mL^{-1} *Escheria coli*. Water for human consumption may not contain any bacteria.

It is important to point out that the population of the Santa Bárbara district (the town itself and the townships Brumal, Barra Feliz and Santana do Morro) did not receive treated water at the beginning of the Project (1998). A portion of the population of Brumal and Barra Feliz started receiving treated water by SAAE (Autonomous Services of Water and Sewage Treatment of Santa Bárbara) as of 2005, while others still had to wait for this service. The people of Santana do Morro used untreated water from the springs until mid-June 2007. The presence of arsenic and coliforms in drinking waters of Santana do Morro justified building a water treatment station within the ARSENEX project, which is now in operation (▶ 16) to meet the necessary quality criteria and deliver clean drinking water to the residents.

Most of the Nova Lima district population received treated water. The investigation that follows refers to water from the springs and ponds that serve the townships of Brumal, Barra Feliz and Santana do Morro of the Santa Bárbara district, part of the Piracicaba river basin. To assess the spring water quality in the Santa Bárbara district, analyses of various physical-chemical and microbiological parameters were carried out (Table 11.4), and results compared with drinking water standards (▶ 5).

With some exceptions, and in agreement with previous observations (above), the pH-values were mostly unproblematic. Individual low pH-values, most likely due to acid mine drainage, need to be monitored. Water colour often leaves much to be desired, frequently related to iron and manganese and in the rainy season to suspended solids. This result is corroborated by the data on dissolved and total solids. Since the regional Fe and Mn-rich soils and bedrock are the source for elevated Fe and Mn-concentrations in water, there is not much that can be done to prevent these enrichments, except for a reduction of soil erosion. As well, the variability of dissolved oxygen is very high and certainly related, jointly with the just as variable BOD-values, to the irregular discharge of mainly human and animal refuse into the waters.

Table 11.3.   Results of total As determinations in spring waters, Santa Bárbara district 2002 and 2005 (this work).

Sample location	Total As (μg L^{-1}) range
Brumal	3.12–15.67
Barra Feliz	2.18–14.37
Santana do Morro	3.26–21.45

Table 11.4. Physico-chemical and microbiological results of spring water in the Santa Bárbara district (this work).

Parameter	Range	Threshold values
pH	5.9–7.9	6–9.5
Colour (mg Pt L^{-1})	<1–181	15
Turbidity (NTU)	0.08–14	5
Suspended solids (mg L^{-1})	<0.2–1.5	–
Dissolved solids (mg L^{-1})	8.1–32.7	500
Total solids (mg L^{-1})	3.6–218	–
Dissolved O_2 (mg L^{-1})	<0.2–10.5	–
BOD (mg L^{-1})	<0.2–11.2	–
NH_4-N (mg L^{-1})	<0.01–0.52	1.5
Soluble Fe (mg L^{-1})	<0.01–1.15	–
Total Mn (mg L^{-1})	<0.01–0.25	0.1
Total Cu (mg L^{-1})	<0.02–0.014	2
Ag (mg L^{-1})	<0.01	–
PO_4^- (mg L^{-1})	<0.01–0.32	–
Cl^- (mg L^{-1})	<0.01–2.1	250
Oils and grease (mg L^{-1})	<1–4.5	–
Fecal coliforms (UFC 100 mL^{-1})	0–2.400	absent
Total coliforms (UFC 100 mL^{-1})	0–5.500	absent
Fecal streptococcus (UFC 100 mL^{-1})	0–360	absent

The more serious quality problems are again related to the hygienic conditions of the waters, as the results obtained for fecal coliforms, total coliforms and fecal streptococcus show (Table 11.4). The ratio between fecal coliforms and fecal streptococcus concentrations allows the origin of the contamination to be assessed, using an empirical criterion (Sperling 1996):

- if CF/EF > 4 – the contamination is predominantly human,
- if CF/EF < 1 – the contamination is predominantly due to other warm-blooded animals,
- if CF/EF is between 1 and 4 – the interpretation is dubious.

Based on the minimum CF/EF values (CF/EF = 0/0 = 0) and the maximum values found (CF/EF = 2400/360 = 6.7) within the study period, the contamination varies especially due to the action of warm-blooded mammals. Therefore, the quality of the waters collected from the springs in the project study area within the study period, cannot be considered satisfactory in terms of potability.

REFERENCES

Bundschuh, J., Pérez Carrera, A. & Litter, M. (eds): *Iberoarsen. Distribución del arsénico en las regiones Ibérica e Iberoamericana.* CYTED Argentina, 2008, p.230. ISBN 13978-84-96023-61-1.

Bundschuh, J., Armienta, M.A., Birkle, P., Bhattacharya, P., Matschullat, J. & Mukherjee, A.B. (eds): *Natural arsenic in groundwaters of Latin America.* In: Bundschuh, J. & Bhattacharya, P. (ser eds), Volume 1. CRC Press/Balkema, Leiden, 2009, p.742.

Chappell, W.R., Abernathy, C.O., Calderon, R.L. & Thomas, D.J. (eds): *Arsenic exposure and health effects*, Volume V. Elsevier, Amsterdam, 2003, p.533.

COPAM: Deliberação normativa COPAM n. 10 from December 16, 1086. Estabelece normas e padrões para qualidade das águas, lançamento de efluentes nas coleções de águas, e dá outras providências (1986) www.novaambi.com.br/pdfs/copam.pdf.

Goyer, R.A., Aposhian, H.V., Brown, K.G., Kantor, K.P., Carlson, G.P., Cullen, W.R., Daston, G.P., Fowler, B.A., Klaasen, C.D., Kosnett, M.J., Mertz, W., Preston, R.J., Ryan, L.M., Smith, A.H., Vather, M.E. & Wienke, J.K.: *Arsenic in drinking water*. National Academy Press, Washington, DC, 1999, p.330. ISBN-10: 0-309-06333-7.

IGAM: Monitoramento das águas superficiais no Estado de Minas Gerais. Projeto "Águas de Minas", 1998–2004, 2008, Instituto Mineiro de Gestão das Águas, Belo Horizonte.

Rawlins, B.G., Williams, T.M., Breward, N., Ferpozzi, L., Figueiredo, B.F. & Borba, R.P.: Preliminary investigation of mining-related arsenic contamination in the provinces of Mendoza and San Juan (Argentina) and Minas Gerais State (Brazil). In: *British Geological Survey Technical Report WC/97/60*, Kenworth, UK, 1997, p.25.

Reimann, C. & de Caritat, P.: *Chemical elements in the environment. Factsheets for the geochemist and environmental scientist*. Springer Verlag, New York, Berlin, 1998, p.398.

Sperling, M. von: Comparison among the most frequently used systems for wastewater treatment in developing countries. *Water Sci. Technol.* 33:3 (1996), pp.59–72.

Vasconselos, O., Oberdá, S., Deschamps, E. & Matschullat, J.: Água. In: Deschamps, E. & Matschullat, J. (eds): Arsênio antropogênico e natural. Um estudo em regiões do Quadrilátero Ferrífero. FEAM, Belo Horizonte, 2007, pp.189–199.

Williams, M.: Arsenic in mine waters: an international study. *Environ. Geol.* 40:3 (2001), pp.267–278.

CHAPTER 12

Soils and sediments

Eleonora Deschamps, Jaime Mello & Jörg Matschullat

There is much debate about what would be natural in terms of concentration of an element in soils or sediments (Reimann *et al.* 2009). Related answers are crucial when health hazards for humans and environment are under scrutiny. Knowledge of the natural As presence is needed to assess environmental quality and to develop actions capable of curbing soil and sediment pollution (▶ 17). An attempt to understand the As biogeochemistry in subtropical conditions led to a series of studies of the Iron Quadrangle that revealed positive As anomalies in different environmental compartments (Borba *et al.* 2000, 2003; Eleutério 1997; Haßler 2002; Matschullat *et al.* 2000; Rawlins *et al.* 1997). The dimension and number of As anomalies are a serious concern in terms of environmental impact and particularly in respect to their effect on public health. A preliminary human biomonitoring study in regions of the Iron Quadrangle revealed that As assimilation by the human body occurred (Matschullat *et al.* 2000; ▶ 14). Based on these results, this chapter presents a more detailed study of surface soil material and evaluates the hypothesis that airborne dust could be decisive in As transfer to the human body (▶ 10). With the analysis of sediment samples, the study seeks to reply to additional questions on As dispersion in the study area. Bearing in mind the development of remedial actions for contaminated soils and sediments, this chapter also contributes to As geochemistry and As mobility in the environment (▶ 1).

Earlier geochemical studies in the Iron Quadrangle generally served mineral prospection rather than contributing to environmental issues (e.g., Oliveira *et al.* 1979). That work included a geochemical map which showed that 76% of 1,297 samples yielded As concentrations below 20 mg kg^{-1}, 20.3% of the samples indicated As values between 20 and 100 mg kg^{-1}, and only 3.7% (48 samples) presented As concentrations from 101 to 5,000 mg kg^{-1}. All materials had been collected in the Nova Lima group (▶ 7), a rock series in an area of active gold mining over the past 300 years. Studies that focused more on environmental questions, including other compartments were published only recently and suggested the need for further studies (Borba *et al.* 2000, 2003; Deschamps *et al.* 2002; Deschamps and Matschullat 2007; Eleutério 1997; Matschullat *et al.* 2000, 2007; Rawlins *et al.* 1997).

While present mining activities in the area appear not to contribute significantly to soil and sediment contamination, potential As intoxication hazards remain, induced by dispersion of old residues, by human occupation of polluted soils and their elevated to very high As levels and by the use of contaminated water – surface or underground (▶ 11).

The As enrichment in soil can affect every environmental compartment, including human health. Considering the relatively low mean value of As concentrations in most natural environments (Matschullat 2000; ▶ 1) and the sensitivity of several biota to this element (▶ 1, 2, 13, 14), many countries introduced low As limit values for soils and sediments (▶ 5). Brazil, similar to the United States of America, permits relatively high soil As values (Table 12.1).

The main sources of As concentrations in soils – and subsequently in sediments – are bedrock materials, pedogenesis and human activities. Pedogenesis defines the behavior of the mineral components from different bedrock. The organic and inorganic soil components, hydrology, redox state and prevalent pH-values affect the element concentrations in the soil. Bedrock composition is far more important than the soil type because it more strongly affects the metal content in the soil (Tang 1985, 1987). Comparatively low humus content, hard to

Table 12.1.   National threshold As values for soils.

	BR	CAN	DE	DK	NL	USA
mg kg^{-1}	15–100	19	5–50	10	34–50	10–100

BR: CETESB (2001), EMBRAPA (1999); CAN: CCME (1997), DE: Throl (2000), DK: Jensen *et al.* (1997); NL: Crommentuijn *et al.* (1997), USA: Efroymson *et al.* (1997).

very hard surfaces due to insolation, excessive and deep-reaching weathering, and yellowish or reddish colours characterize the prevalent oxisols of the Iron Quadrangle with its tropical and subtropical climate (▶ 7). The hard surface crusts erode mechanically, producing a very fine dust of oxic powder. This powder is transported by wind and by rainwater and sheet flow. As a result, this material feeds the local drainage system and generates much of its sediments. At the same time, soils are receptors of atmospheric deposition and may bear airborne particles from further distances (▶ 10). The resulting soil geochemistry delivers signals of a variable mixture of these processes and, where applicable, local contamination. Consequently, the fine dust – the material of the immediate pedosphere-biosphere interface – is the tool that allows evaluating geochemical anomalies and discerning residential and leisure areas (particularly for children risk assessment).

12.1   SAMPLING OF SOILS AND SEDIMENTS

Soils. Composite samples of this surface soil material were collected on areas of at least 10,000 m^2 in the Nova Lima and Santa Bárbara districts between 1998 and 2003. The material was gathered with a stainless steel blade (scraping hard surfaces) or HDPE-blades (Figure 12.1). During sampling, coarse mineral compounds and plant material (such as roots etc.) were actively avoided. Each sample (150 ± 50 g) was placed in 180 mL PE-bags (WhirlPack®), labeled and sealed after air-drying in the field. Additional freeze-drying to constant weight was performed later in the laboratory (Alpha I, Christ, Germany). After dry sieving (63 μm nylon mesh), the fine material (usually >90%) was used for the subsequent analyses, while the coarser material (>63 μm) was discarded.

   Sediments. Surface sediments from rivers and creeks were treated similar to the soil material. Given the large interannual hydrological variations (discharge with level variations of more than 5 m), sediment accumulation and dispersion are much more dynamic in tropical environments as compared to moderate climate zones. A single sediment sampling campaign may not yield reliable information in respect to distribution and accumulation of compounds. Therefore, several sampling campaigns were carried out in different seasons (April and August). The sediments were collected manually in brooks and streams during low-flow times. During sampling, the more coarse sediments were discarded. Non-consolidated fresh material was collected in PE flasks from the bentonic surface, and each sample (around 500 mL) represented an area of about 100 m^2 of surface sediments. Most of the samples were freeze-dried to constant weight, and then sieved as above.

12.2   PREPARATION AND ANALYTICAL METHODS

For maximal quality control and to gather additional experience with soil and iron-rich sediment matrixes, parallel analyses of the samples with different and independent methods was planned from the beginning. Sub-samples were used: a) directly for solid sampling analysis

Figure 12.1. Soil sampling with stainless steel shovel (▶ 12; photo Michelly Amorim).

by atomic absorption spectrometry (SS-GF-AAS), b) after total acid digestion for analysis by inductively-coupled plasma – quadrupol mass spectrometry (ICP-QMS), and c) by energy dispersive (ED) and wave-length-dispersive (WD) X-ray fluorescence spectrometry (XRF) analyses on pressed powder pellets for the independent determination of major, minor, and selected trace elements (Schmidt 2001). Individual samples were also tested by instrumental neutron activation analysis (INAA), particularly to verify the accuracy of difficult As analysis in complex matrices (e.g., soils and sediments).

No further sample preparation was needed for the SS-GF-AAS analysis; the fine material was used directly. For ICP-QMS analysis, the samples were digested in a microwave oven (MWS 1, Maassen, Germany) in 80 mL PTFE recipients. The temperature was kept at 175°C for 20 min with a pressure of ca. 25 bar (2.5×10^6 Pa). After cooling, the samples were transferred to 25 mL and 50 mL HDPE-flasks, respectively, which were filled with de-ionized water. For XRF (ED and WD) analysis, ca. 5 g of sample powder were homogenized and pressed under 20 t for 30 seconds with a bonding agent (Hoechst wax). Pyrolysis at 400°C and atomization at 2,100°C led to a precise and reproducible As determination at 193.7 nm by SS-GF-AAS (5EA, Analytik Jena, Germany). Standard As solutions were used in the calibration (Merck, Germany). A comparison with the certified reference material (STSD-2, Canada), found 40 ± 1.5 mg As kg^{-1} (6 replicates) instead of 42 mg As kg^{-1} (recommended value).

Calibration for the ICP-QMS analysis (PE 9000; Perkin Elemer Elan) was carried out with reference solutions containing 1, 20, 50, and 100 µg As L^{-1}, prepared with a stock solution of 1 mg As L^{-1} (ICP-MS standard, Merck) by dilution of 1% in nitric acid. Rhodium (Rh) was used as internal standard (ICP-MS standard, 1 g L^{-1}, Merck). Rh concentration was 10 µg L^{-1}, both in the samples and the reference solutions. To acidify and digest samples, the following chemicals were used: nitric acid (Suprapur, Merck), hydrogen peroxide (30%, Suprapur, Merck) and hydrochloric acid (Instra-Analyzed, Baker). The certified reference material NIST 1643d "Trace-elements in Water" was measured on a regular basis for quality control. Errors remained below 5% RSD. The WD-XRF analyses (PW 2480) were carried out with the program UniQuant (Philips, Holland) and with the certified reference materials as unknown samples. ED-XRF was run only on selected samples for additional quality control. Low As values led to positive errors of 21% RSD (<100 mg kg^{-1}), while higher As values of all its other compounds yielded values below 3% RSD.

12.3   SOIL AND SEDIMENT RESULTS, HEALTH RISK ASSESSMENT

Soils. A simplified view on the encountered As anomalies (Figure 12.2) and on soil geochemistry (Table 12.2) in the Nova Lima and Santa Bárbara districts, shows a wide variation of mean values, untypically high for the region, and a distinct difference between the two districts. These results are compatible with other data (Ladeira 1999). The relative enrichment of Al and Fe and the depletion of Ca, Na and Mg (Table 12.2) clearly reveal their origin from the tropical and subtropical soils of Minas Gerais. With the exception of Zn and Pb, and to a certain extent Cu, the trace-elements As, Co, Cr and Ni show enrichments, relative to the global average soil concentrations. This reflects both a lithologic situation in the Iron Quadrangle, in respect to arsenic, and the additional As dissipation from centuries of mining and smelting activities.

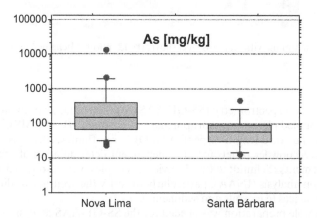

Figure 12.2.   Arsenic concentrations in soils of the Nova Lima and Santa Bárbara districts.

Table 12.2.   Rounded mean values and concentration ranges of composite soil samples from the Iron Quadrangle, compared with the average values of world soils (WSA)*. Major and minor compounds in wt.-%, trace-elements in mg kg^{-1}.

	Nova Lima n = 21	Santa Bárbara n = 13	WSA*
SiO$_2$	59 (49.7–68.1)	39 (12.9–59.4)	60
Al$_2$O$_3$	24 (19.3–28.6)	28 (17.8–36.4)	15
Fe$_2$O$_3$	13 (3.2–21.0)	27 (7.8–54.7)	5
CaO	0.9 (0.01–3.8)	1.1 (0.09–5.1)	1.9
Na$_2$O	0.2 (0.03–1.0)	0.3 (<0.03–0.74)	1.3
K$_2$O	2.4 (1.14–3.39)	1.8 (0.39–3.44)	1.7
MnO	0.09 (0.02–0.18)	0.16 (0.01–0.39)	0.07
MgO	0.6 (0.30–1.50)	0.6 (0.28–1.24)	1.5
TiO$_2$	0.9 (0.56–1.22)	1.2 (0.69–1.66)	0.7
As	960 (16–13,400)	100 (13–467)	5
Co	71 (13–102)	117 (32–160)	10
Cr	410 (190–680)	435 (150–730)	80
Cu	44 (22–90)	53 (18–92)	25
Ni	163 (63–300)	93 (20–220)	20
Pb	23 (11–32)	33 (19–51)	17
Zn	71 (12–240)	93 (38–170)	70

*Compilation from Reimann and Caritat (1998).

Natural As concentrations in soils and rocks are generally very low (Matschullat 2000; Riedel and Eikmann 1986; Tanaka 1988; ▶ 1). Elevated As concentrations occur in magmatic sulphides and iron ore. Related soils generally exceed the normal As levels (Huang 1994). In regions with volcanic activity (e.g., Chile, Italy, Japan, Mexico), the mean As content in soils may reach and exceed 20 mg kg⁻¹, while "normal" soils present an average of 2 mg kg⁻¹ (Vinogradov 1959). Reimann *et al.* (2009) show the enormous natural variability with examples from Europe that mainly reflect geological (and soil formation) processes. As species in the Iron Quadrangle occur in primary gold mineralizations as discrete mineral phases, such as arsenopyrite and loellingite, or as impurities in remaining sulphides, especially pyrite (Borba *et al.* 2003). The As/Au ratio in these minerals ranges from 300 to 3,000 between deposits.

Figures 12.3 and 12.4 show the sampling sites and their soil As concentrations. The minimum values were below 10 mg kg⁻¹ (green) and represent the local situation. As concentrations between 10 and 50 mg kg⁻¹ (blue) occur far (>1 km) from gold mining and ore processing areas and can be regarded as natural anomalies, while concentrations >100 mg kg⁻¹ (yellow) were exclusively found near larger gold mines and mineralizations. Concentrations >1000 mg kg⁻¹ (red) normally occurred only close to or on mine residue deposits (e.g., tailings). The main difference between the samples collected in Nova Lima and Santa Bárbara can be explained by the long years of mining activity and smelting in Nova Lima, while these activities were

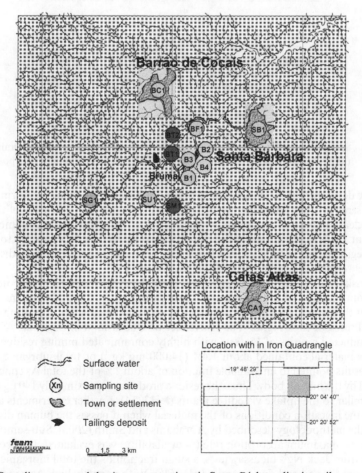

Figure 12.3. Sampling spots and the As concentrations in Santa Bárbara district soils.

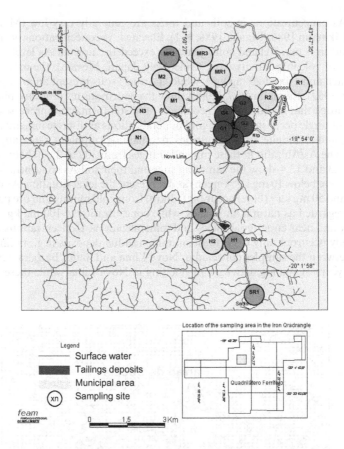

Figure 12.4.   Sampling spots and the As concentrations in Nova Lima district soils. Drainage network (lines), settlements (light areas), reservoirs (dark bodies), and tailings deposits.

and still are developed on a smaller scale in Santa Bárbara. These results confirm the hypothesis that arsenic in rocks and minerals oxidates during weathering, and some of the arsenic is naturally released in the environment, while mining activities increase the As mobility in the environment. Soils in Nova Lima were characterized by an enrichment of up to twenty (20) times the regional background value, while the samples of the Santa Bárbara district present a moderately enriched mean concentration with 50 mg kg^{-1}.

Health risks. Additional experiments were carried out to better estimate potential health hazards from human soil uptake, since As concentrations in soil and dust are not equal to the amount that is potentially sorbed by the (human) organism after inhalation or ingestion (▶ 2, 14). Professor Jack Ng, of The University of Queensland in Brisbane, Australia, generously conducted a series of biotests with highly contaminated mining residues from Galo Novo. This material contained about 1.4% (14,000 mg kg^{-1}) of total inorganic arsenic. His bioassay results pointed to an available fraction of about 1% of the total As concentration to be retained by the human body. Thus, the residue produces approximately 140 µg of As g^{-1}, an amount similar to the uptake via edible plants (▶ 13, 14). Further experiments attempted to reproduce the ingestion conditions of the material when it passes the human digestive tract, following the methodology described by Abrahams and Smith (2003). Sub-samples were collected and the maximum absorption rate (= availability) was evaluated. In accordance with the results from Jack Ng's bioassay tests, a small fraction of the total inorganic arsenic was available from the soil and dust samples (<1–15%). With partially high As concentrations in the material, the contribution of this fraction to total uptake again is in the same magnitude as through As-enriched food. Table 11.3 shows the absolute ingestion by younger children, based on a regular mean ingestion of 500 mg of soil material. While no relevant contribution was found in the samples from Santa Bárbara, a substantial addition to the daily ingestion was found in the samples from Nova Lima.

People in rural areas of the Iron Quadrangle are – irrespective of age or gender – exposed to environmental dusts. Their homes usually do not have windows and people are adapted to a more open-air lifestyle. Direct inhalation of dusts is another As uptake pathway for these people (▶ 9). Since dust may be As-laden from soil particles, it was also deemed necessary to evaluate related risks. Considering the mean As concentration in the soil, adults (with a breathing volume of 20 m^3 day^{-1}) may ingest 1.5 µg As a day in Santa Bárbara and 6 µg As a day in Nova Lima, while children with a breathing volume of 5 m^3 day^{-1} would inhale between 0.4 µg As a day in Santa Bárbara and 1.5 µg As a day in Nova Lima. These amounts would reach the most sensitive parts of their lungs, the alveoli, due the small grain size (<63 µm). Aiming at reducing this exposure, the project organized environmental education workshops to encourage people to reduce domestic dust exposure as much as possible (▶ 15).

Sediments. From an environmental point of view, fine grained sediments (<63 µm) are a reliable tool to evaluate the environmental quality of a basin (Salomons and Förstner 1984). Due to their large surface, this material has high absorption capacities – and it is rather homogeneous. The sediment samples from the districts of Nova Lima and Santa Bárbara yielded a rounded mean value of 45% <63 µm (range 7–77%) and 55% >63 µm (23–93%).

The highest As concentrations occurred with a mean value of 190 mg kg^{-1} in the Nova Lima district, while a mean value of 86 mg kg^{-1} was detected in the Santa Bárbara district (Table 12.4). The maximum values in Nova Lima were found close to large residue deposits (tailings) and in secondary streams that drain the mining and ore processing areas. The highest concentrations occurred in sediments of the Cardoso stream that flows through the residue deposit in the Galo community (▶ 7, 11). There, As concentrations between 47 and 3,300 mg kg^{-1} (mean 547 mg kg^{-1}) were measured.

The highest As concentrations in Santa Bárbara occurred close to the São Jorge mine near Brumal. That place consists of an excavated area of approximately 100 m (L) × 50 m (W) × 30 m (D) in rocks of the Nova Lima group (▶ 7.1). Apparently, the Au-production comes from a ferruginous quartzite and a quartz vein, containing pyrite, pyrrhotite and arsenopyrite, and native gold in very fine grains (Prado *et al.* 1991). Arsenic, not retained by Fe-Mn oxihydroxides, is transported to the rivers with the surface drainage.

The concentrations show major variations along the river courses and in between sampling campaigns in both districts. While the mean values remain almost identical for an extended period of time – indicating constant input – individual sampling locations may show differences in magnitude. In Rio das Velhas, samples from April 1999 had 3,300 mg As kg^{-1} before its confluence with the Cardoso stream; the same sampling site yielded 100 mg As kg^{-1} in August 1998. The Cardoso stream, above the confluence with Rio das Velhas, presented

Table 12.3. Daily As ingestion and As amounts reabsorbed from soil via hand-to-mouth contact in children aged 1 to 6 years (500 mg of soil per day).

District	Township	Soil As (mg kg^{-1})	Daily ingestion (µg As)	Daily resorption (µg As)
Santa Bárbara	Brumal	42	20	<<1
	Barra Feliz	83	40	<<1
Nova Lima	Galo Novo	360	180	20
	Galo Velho	165	85	10
	Matadouro	1600	180	20

Table 12.4.   Sediment geochemistry in the Iron Quadrangle compared with mean world sediment*. Major and minor compounds in wt.–%, trace elements in mg kg^{-1}.

	Nova Lima n = 24	Santa Bárbara n = 18	Average global sediment*
SiO_2	42.6–61.8	37.7–65.0	61.3–63.8
Al_2O_3	10.8–26.2	6.6–19.3	1.2–16.6
Fe_2O_3	8.8–23.7	10.9–59.3	1.4–7.8
CaO	0.16–0.50	0.06–0.32	0.57–1.80
Na_2O	0.16–0.26	0.19–0.25	0.04–2.0
K_2O	1.2–3.6	0.48–0.86	0.17–2.78
MnO	0.07–0.57	0.10–0.37	0.07–0.25
MgO	0.67–1.2	0.4–7.1	0.63–2.39
TiO_2	0.58–1.02	0.45–1.35	0.21–0.97
As	47–3,300	22–160	2–22
Co	11–62	3–86	7–32
Cr	24–640	17–1,150	24–161
Cu	36–82	26–93	12–24
Ni	31–190	15–510	14–59
Pb	9–57	7–48	8–195
Zn	46–130	27–160	44–209

*From Reimann and Caritat (1998).

similar changes from 120 mg As kg^{-1} in April 1998 to 2,400 mg As kg^{-1} in August 1999. Similar variations occur in other rivers and streams as a result of the redistribution of sediments during the wet season (high discharge and flooding) and the dry season (low discharge).

Most of the surface soil and sediment samples exceeded the global mean values and all values recommended by national and international bodies (Tables 12.1 and 12.4). This reflects the distribution of positive As anomalies in the Iron Quadrangle, related with lithology and with historical and current mining and smelting activities.

12.4   ARSENIC – GEOCHEMISTRY AND ENVIRONMENTAL MOBILITY

Laboratory studies showed that both arsenate, $As^{(V)}$, and arsenite, $As^{(III)}$, are sorbed to Fe and Al oxihydroxides (▶ 3, 4). This suggests a similar behaviour in the environment. The high affinity of ferrous oxihydroxides with $As^{(V)}$ is well known, however, $As^{(III)}$ also seems to specifically sorb to these oxides. Both can form binuclear "inner-sphere" complexes with the hydroxilated surfaces of the ferrous oxide, as shown by infrared spectroscopy (FTIR; Sun and Doner 1996) and by X-ray absorption spectroscopy (Deschamps et al. 2003; Manning et al. 1998; XAS). However, it has been observed that $As^{(III)}$ is generally more mobile than $As^{(V)}$ in soils and sediments, apparently due to the reductive dissolution of the ferrous oxide matrix, followed by a reduction to arsenate and arsenite.

Meng et al. (2001) demonstrated a notable coincidence in the reduction rates of Fe and liberation of $As^{(III)}$ in the liquid phase of a water treatment suspension, anaerobically incubated for 80 days. Experimental results and modeling studies suggest that the As liberation occurs in an intermediary pE band, between –4 and 0, when transformations between oxidated and reduced solid phases were observed. High S and Fe-content of the sediments can immobilize arsenic because of the formation of arsenopyrite at pE values below –4. These limits may undergo alterations in natural systems, depending on the kinetics of oxidation-reduction reactions and the minerals involved in As immobilization.

Similar results were found by Mello *et al.* (2006a) with samples from Santa Bárbara that were incubated under anaerobic conditions for 120 days (Figure 12.5). The reductive Fe-dissolution rate in these soils was low, and As liberation was limited. Arsenic was released into the solution in one sample, only after the dissolution of Fe and Mn. These results are in accordance with the general concept that ferric oxides have a high As affinity and control its mobility. Under anaerobic conditions, however, the reduction of these Fe-oxides can cause As release into waters. Research in the last decade established that the As mobility in natural environments is mainly controlled by sorption reactions on the surface of Fe and Al oxides. The magnitude of this process, however, can be influenced by other dissolved substances, capable of interacting with the surfaces or arsenic itself. Appelo *et al.* (2002) suggest that dissolved carbonates in groundwater can affect the As sorption to Fe-oxides.

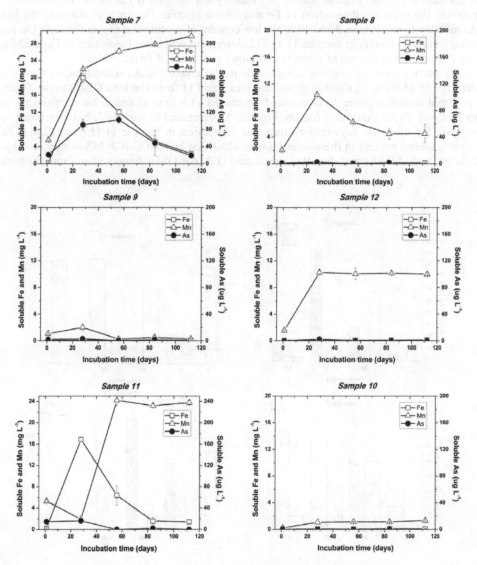

Figure 12.5.   Iron, Mn and As concentrations over time under anaerobic incubation in soil solutions from the Iron Quadrangle (Mello *et al.* 2006a).

Organic matter, with its prevalence in water and on Earth's surface, may affect As mobility due to the high reactivity of some organic compounds with metals. Organic matter can reduce the sorption of both As$^{(III)}$ and As$^{(V)}$ onto the surface of hematite (Redman *et al.* 2002). They also observed the influence of organic matter in the state of oxireduction and in As speciation. Considering that arsenite was consistently more desorbed than the arsenate, the authors suggest that these interactions can be the reason for the greater mobility of As$^{(III)}$ in relation to As$^{(V)}$ in the environment.

The role of organic matter in the dynamics of oxireduction and competitive As sorption was also observed in soils of the Iron Quadrangle (Mello *et al.* 2006a; Figure 12.6). The samples 7 and 8, respectively, were collected from soil surface (0–20 cm) and sub-surface (20–50 cm) of the same profile. Both have the same mineralogy and only differ in the organic carbon (C$_{org}$) content. However, As mobilization was visible only in sample 7, richer in C$_{org}$. It is assumed that the organic matter was used by microorganisms as an electron source to promote the reductive dissolution of Fe and release arsenic. The authors attribute the low As mobility from samples 9 and 10 to the low content of arsenic and C$_{org}$ in the soils. On the other hand, As stability in samples 11 and 12, despite the reductive dissolution of Fe and Mn, was related to the presence of gibbsite and larger quantities of ferric oxides.

The participation of organic compounds in the anaerobic As solubilization was also detected by Mello *et al.* (2006b) in soil samples 7 and 11 from the Iron Quadrangle. There, As$^{(III)}$ and soluble organic As compounds increased with time of anaerobic incubation (up to 28 days). From that point on, the organic As decreased in sample 7, but continued to increase for up to 56 days under anaerobic incubation in sample 11 (Figure 12.6). The organic species present in these samples were identified by HPLC-ICP-MS as monomethylarsonic acid (MMA) and dimethylarsinic acid (DMA); DMA having been found in both

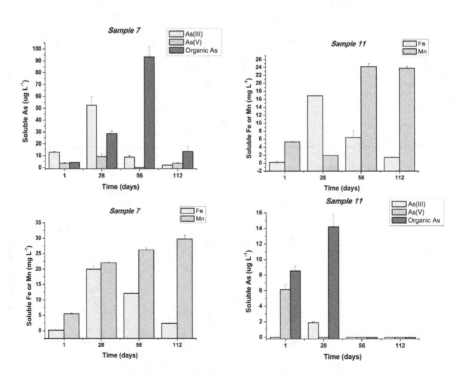

Figure 12.6.   Soluble As species (top) and soluble Fe and Mn (bottom) during anaerobic incubation of soil samples from the Iron Quadrangle (Mello *et al.* 2006b).

samples. The authors found two additional As species, possibly organic, but not identified (see e.g., Planer-Friedrich *et al.* 2006). The studies by Mello *et al.* (2006a, b) revealed that easily exchangeable As, also considered easily "lixiviable", from the soils was mainly $As^{(III)}$, and that $As^{(V)}$ and organic As species were responsible for the increased As mobility, due to the reductive dissolution of Fe and Mn. These results are surprising for anaerobic conditions in terrestrial environments. The unexpected occurrence of organic species in relevant amounts was attributed to the formation of ternary organic complexes or biomethylation by photosynthesizing microorganisms such as algae or cyanobacteria (Planer-Friedrich *et al.* 2006). It is believed that these findings, if proven in loco, can be useful in remediation strategies for contaminated soils and sediments, since most organic As species are considered less toxic than the inorganic species and, in turn, $As^{(V)}$ is considered less toxic than $As^{(III)}$ (▶ 2).

Arsenic speciation in environmental samples deserves increasing attention, since the toxic effects of arsenic and its mobility are related to its oxidation state. Therefore, the individual quantification of the different chemical species of the element is essential when toxicity and biotransformation in aquatic environments are to be studied (▶ 14).

The solid As phases in sediments can be investigated by various methods, including XAS, X-ray diffraction (XRD) and chemical extractions. XAS analyses can yield detailed information about speciation and the As coordination environment in the solid phase, but in many cases this technique is not used due to the limited access to instrumentation and difficulties in interpreting the results of heterogeneous natural samples. XRD-analysis allows identifying the results of mineralogical As-containing phases, but the technique is limited to badly crystallized minerals and small amounts in the solid phase. Because of these limitations, Keon *et al.* (2001) tested a procedure that consists of sequential chemical extractions to identify the As forms in the solid phase: i) slightly and strongly sorbed As, ii) co-precipitated arsenic with oxides or amorphous monosulphide, and iii) arsenic precipitated with crystalline iron, iv) As oxides, v) co-precipitated arsenic with pyrite, and vi) As sulphides.

Santana Filho (2005) used the sequential chemical extraction described by Chunguo and Zihui (1988) to study the distribution of the different As forms in profiles of As contaminated soil in Minas Gerais. The results showed that the most labile forms (and therefore capable of higher mobility and potential bioavailability), make up a very small fraction of the total arsenic. The significant As amounts in soils from the Iron Quadrangle, coupled with organic matter, certainly have an influence on As mobility and availability. These results confirm the importance of the involvement of organic compounds in the As biogeochemistry in this region (Mello *et al.* 2006b). However, that study did not confirm the predominance of arsenic associated with Fe-oxides, as was expected in those soils. The result was possibly caused by some methodological fault. It appears plausible that the adopted fractionation method (Chunguo and Zihui 1988) only takes into account arsenic that is sorbed on the surface of Fe-oxides as "As associated with Fe", while most arsenic remained more strongly bound or co-precipitated onto ferric oxides in the residual phase.

REFERENCES

Abrahams, P.W. & Smith, B.: The geochemical implications of geophagy (deliberate soil consumption) undertaken by Asians residing in the U.K. *6th Internat. Symp. Environ. Geochem.*, Univ of Edinburgh, Scotland; Abstract O70, 2003, p.52.

Appelo, C.A.J., Van Der Weiden, M.J.J., Tournassat, C. & Charlet, L.: Surface complexation of ferrous iron and carbonate on ferrihydrite and the mobilization of arsenic. *Environ. Sci. Technol.* 36:14 (2002), pp.3096–3103.

Borba, R.P., Figueiredo, B.R., Rawlins, B. & Matschullat, J.: Arsenic in water and sediment in the Iron Quadrangle, state of Minas Gerais, Brazil. *Rev. Brasil Geociências* 30:3 (2000), pp.554–557.

Borba, R.P., Figueiredo, B.R. & Matschullat, J.: Geochemical distribution of arsenic in the Iron Quadrangle, State of Minas Gerais, Brazil. *Environ. Geol.* 44:1 (2003), pp.39–52.

Bundschuh, J., Armienta, M.A., Birkle, P., Bhattacharya, P., Matschullat, J. & Mukherjee, A.B. (eds): *Natural arsenic in groundwaters of Latin America.* In: Bundschuh, J. & Bhattacharya, P. (ser eds), Volume 1. CRC Press/Balkema, Leiden, 2009, p.742.

CCME: Recommended Canadian soil quality guidelines. Canadian Council of Ministers of the Environment, Winnipeg, Manitoba, Canada, 1997, pp.1–6.

CETESB: Relatório de Qualidade das águas subterrâneas no Estado de São Paulo 1998–2000. Companhia de Tecnologia de Saneamento Ambiental. São Paulo. *Série Relatórios 96*, 2001, ISSN 0103-4103.

Chappell, W.R., Abernathy, C.O., Calderon, R.L. & Thomas, D.J. (eds): *Arsenic exposure and health effects*, Volume V. Elsevier, Amsterdam, 2003, p.533.

Chunguo, C. & Zihui, L.: Chemical speciation and distribution of arsenic in water, suspended solids and sediment of Xiangjiang River, China. *Sci. Total Environ.* 77:1 (1998), pp.69–82.

Crommentuijn, T., Polder, M.D. & van de Plassche, E.J.: Maximum permissible concentrations and negligible concentrations for metals, taking background concentrations into account. Report No. 601 501 001. Rijksinstituut vor Volksgezondheid en Milieuhygiene (RIVM), Bilthoven, the Netherlands, 1997, p.260.

Deschamps, E., Ciminelli, V.S.T., Lange, F.T., Matschullat, J., Raue, B. & Schmidt, H.: Soil and sediment geochemistry of the Iron Quadrangle, Brazil. The case of arsenic. *J. Soils Sediments* 2:4 (2002), pp.216–222.

Deschamps, E., Ciminelli, V.S.T., Weidler, P.G. & Ramos, A.Y.: Arsenic sorption onto soils enriched with manganese and iron minerals. *Clays Clay Minerals* 5:2 (2003), pp.197–204.

Deschamps, E. & Matschullat, J. (eds): *Arsênio antropogênico e natural. Um estudo em regiões do Quadrilátero Ferrífero.* Fundação Estadual do Meio Ambiente, Belo Horizonte, 2007, p.330. ISBN 978-85-61029-00-5.

Efroymson, R.A., Will, M.E. & Suter II, G.W.: Toxicological benchmarks for contaminants of potential concern for effects on soil and litter invertebrates and heterotrophic process. East Tennessee Technology Park Technical Information Office ES/ER/TM-126/R2 report for US DOE, 1997, p.151.

Eleutério, L.: *Diagnóstico da situação ambiental da cabeceira da bacia do rio Doce, no âmbito das contaminações por metais pesados em sedimentos de fundo.* Unpubl. Master thesis, Dept. of Geology, Federal Univ Ouro Preto, Minas Gerais, 1997, p.163.

EMBRAPA: *Sistema Brasileiro de classificação de solos.* Brasilia, SPI, 1999, p.492.

Haßler, S.: *Geochemie von Böden und deren Wechselwirkung mit Pflanzen im südlichen Eisernen Viereck, Brasilien.* Unpubl. M.Sc. thesis, TU Bergakademie Freiberg, 2002, p.124 + annex.

Huang, Y.C.: Arsenic distribution in soils. In: Nriagu, J.O. (ed): *Arsenic in the environment. Advances in environmental science and technology*, Volume 26:I. Wiley, New York, 1994, pp.18–49.

Jensen, J., Lakkenborg-Kristensen, H. & Scott-Fordsmand, J.J.: Soil quality criteria for selected compounds. Danish Environmental Protection Agency Working Report, 83, 1997, p.134.

Keon, N.E., Swartz, C.H., Brabander, D.J., Harvey, C. & Hemond, H.F.: Validation of an arsenic sequential extraction method for evaluating mobility in sediments. *Environ. Sci. Technol.* 35 (2001), pp.2778–2748.

Ladeira, A.C.Q.: *Utilização de solos e minerais para imobilização de arsênio e mecanismo de adsorção.* PhD thesis in Metallurgical Engineering, Federal University of Minas Gerais, Belo Horizonte, Brazil, 1999.

Manning, B.A., Fendorf, S.E. & Goldberg, S.: Surface structures and stability of arsenic(III) on goethite: Spectroscopic evidence for inner-sphere complexes. *Environ. Sci. Technol.* 32:16 (1998), pp.2383–2388.

Matschullat, J.: Arsenic in the geosphere – a review. *Sci. Total Environ.* 249:1–3 (2000), pp.297–312.

Matschullat, J., Borba, R.P., Deschamps, E., Figueiredo, B.R., Gabrio, T. & Schwenk, M.: Human and environmental contamination in the Iron Quadrangle, Brazil. *Appl. Geochem.* 15:2 (2000), pp.181–190.

Matschullat, J., Birmann, K., Borba, R.P., Ciminelli, V.S.T., Deschamps, E.M., Figueiredo, B.R., Gabrio, T., Haßler, S., Hilscher, A., Junghänel, I., de Oliveira, N., Schmidt, H., Schwenk, M., de Oliveira Vilhena, M.J. & Weidner, U.: Long-term environmental impact of As-dispersion in Minas Gerais, Brazil. In: Bhattacharya, P., Mukherjee, A.B., Bundschuh, J., Zevenhoven, R. & Loeppert, R.H. (eds): *Arsenic in soil and groundwater environments: biogeochemical interactions.* Trace metals and other contaminants in the environment, Volume 9. Elsevier, Amsterdam, 2007, pp.365–382.

Mello, J.W.V. de. Roy, W.R., Talbott, J.L. & Stucki, J.W.: Mineralogy and arsenic mobility in arsenic-rich Brazilian soils and sediments. *J. Soils Sediments* 6:1 (2006a), pp.9–19.

Mello, J.W.V. de. Talbott, J.L., Scott, J., Roy, W. & Stucki, J.W.: Arsenic speciation in arsenic-rich Brazilian soils from gold mining sites under anaerobic incubation. *Environ. Sci. Pollut. Res.* 14:6 (2006a), pp.388–396.

Meng, X., Korfiatis, G.P., Jing, C. & Christodoulatus, C.: Redox transformations of arsenic and iron in water treatment sludge during aging and TCLP extraction. *Environ. Sci. Technol.* 35:17 (2001), pp.3476–3481.

Oliveira, J.J.C., Ribeiro, J.H. & Souza, H.Á.: *Projeto Geoquímica do Quadrilátero ferrífero: levantamento orientativo e regional.* Relatório final, Volume VI. CPRM, Belo Horizonte, 1979, p.132.

Planer-Friedrich, B., Ball, J.W., Matschullat, J. & Nordstrom, D.K.: Volatile arsenic and other volatile metal(loids) at Yellowstone National Park. *Geochim. Cosmochim. Acta* 70:10 (2006), pp.2480–2491.

Prado, M.G.B., Pereira, S.L.M., Rodrigues, J.L.L. & Ribeiro, P.A.: Synthesis of the geology of São Bento and Santa Quitéria mines, Santa Bárbara, Minas Gerais. In Fleischer, R., Grossi Sad, J.H., Fuzikawa, K. & Ladeira, E.A. (eds): *Field and mine trip to Quadrilátero Ferrífero, Minas Gerais, Brazil: Field guidebook of Brazil Gold* '91, Internat Symp Geology of gold, Belo Horizonte, Balkema, Rotterdam, 1991, p.103.

Rawlins, B.G., Williams, T.M., Breward, N., Ferpozzi, L., Figueiredo, B.F. & Borba, R.P.: Preliminary investigation of mining-related arsenic contamination in the provinces of Mendoza and San Juan (Argentina) and Minas Gerais State (Brazil). In: British Geological Survey Technical Report WC/97/60, Kenworth, UK, 1997, p.25.

Redman, A.D., Macalady, D.L. & Ahmann, D.: Natural organic matter affects arsenic speciation and sorption onto goethite. *Environ. Sci. Technol.* 36:13 (2002), pp.2889–2896.

Reimann, C. & de Caritat, P.: *Chemical elements in the environment. Factsheets for the geochemist and environmental scientist.* Springer Verlag, New York, Berlin, 1998, p.398.

Reimann, C., Matschullat, J., Birke, M. & Salminen, R.: Arsenic distribution in the environment: the effects of scale. *Appl. Geochem.* 24:7 (2009), pp.1147–1167.

Riedel, F.N. & Eikmann, T.: Natural occurrence of arsenic and its compunds in soils and rocks. *Wiss. Umwelt* 3–4 (1986), pp.108–117.

Salomons, W. & Förstner, U.: Metals in the hydrocycle. Springer, 1984, p.349.

Santana Filho, S.: *Distribuição de arsênio em solos e sedimentos e oxidação de materiais sulfetados de áreas de mineração de ouro do estado de Minas Gerais.* PhD thesis, Federal University of Viçosa, Minas Gerais, Brazil, 2005.

Schmidt, H.: *Arsenbelastung von Oberflächenproben subtropischer Böden aus dem Eisernen Viereck, Minas Gerais, Brasilien.* Unpubl. M.Sc. thesis, TU Bergakademie Freiberg, IÖZ, 2005, p.118.

Sun, X. & Doner, H.E.: An investigation of arsenate and arsenite bonding structures on goethite by FTIR. *Soil Sci.* 161:12 (1996), pp.865–872.

Tanaka, T.: Distribution of arsenic in the natural environment with emphasis on rocks and soil. *Appl. Organomet.* Chem. 2:4 (1988), pp.284–295.

Tang, S.: Study on the relationship between heavy metal in soil a parent material principal component analysis. *Huanjing* 6:3 (1985), pp.2–6.

Tang, S.: Factors affecting the geochemical background contents of heavy metals in different soils of China. *Acta. Sci. Circumstantiale* 7:3 (1987), pp.245–252.

Throl, C.: Ableitung ökotoxikologisch begründeter Bodenqualitätskriterien am Beispiel Arsen. *Z. Umweltchem. Ökotox.* 12:3 (2000), pp.137–148.

Vinogradov, A.P.: *The geochemistry of rare and dispersed chemical elements in soils.* 2nd ed.: Consultants Bureau, New York, 1959, pp.654–70.

CHAPTER 13

Arsenic in edible and bioaccumulating plants

Olívia Vasconcelos, Helena Palmieri, Jörg Matschullat &
Eleonora Deschamps

This chapter deals with the incorporation of arsenic in edible plants and in some native fern species in the Iron Quadrangle. Two major objectives motivated this approach: a) to evaluate the As exposure of the population of the districts of Santa Bárbara and Nova Lima through the ingestion of vegetables and herbs cultivated in their backyards and in As-contaminated areas; and b) to conduct a preliminary study of the As phytoremediation capacity of some native fern species in the municipality of Mariana, Minas Gerais, with the potential to be used to recover As-contaminated areas.

13.1 ARSENIC UPTAKE IN PLANTS

As with most elements, the extent of As uptake by plants from the soil varies between the different plant species. Unlike some marine and freshwater organisms with high concentrations (>1000 mg As kg^{-1} fresh weight in some macrophytes, similar to the related sediments concentration), the level of As uptake in terrestrial plants remains below the level found in the soil (Matschullat 2000; O'Neill 1995; ▶ 1). It is not common to find As concentrations above 1 mg kg^{-1} fresh weight in natural terrestrial plants (Dudka and Miller 1999). Above this level, As-related phytotoxic effects emerge (O'Neill 1995). In general, roots contain higher As concentrations than stems, leaves or fruit. As a result, plants generally do not accumulate enough arsenic to be hazardous to humans in non-contaminated environments. In contaminated soils, however, the As uptake into plant tissue may be significantly higher, particularly in vegetables and edible plants (Table 13.1).

Larsen *et al.* (1992) found significant As levels in vegetables and grains cultivated in As-contaminated soils. Significant As quantities were also found in carrots (*Daucus carota* L.) and other vegetables cultivated in contaminated soil (Helgesen and Larsen 1998; Pyles and Woolson 1982). Grass species, collected in reject piles from mines, were found to have

Table 13.1. Mean As concentrations (mg kg^{-1} dry mass) in plants cultivated in non-contaminated and contaminated soil (Fiedler and Rösler 1987; Matschullat 2000).

Plant	As concentration in plants (mg kg^{-1} dry weight)		Soil As concentration
	Non-contaminated soil	Contaminated soil	(mg kg^{-1} dry weight)
Carrot	0.040–0.080	0.11	17
		0.54	217
		1.2	420
		2.5	650
Lettuce	0.020–0.250	1.6	17
		11	420
		14	650

concentrations of up to 3,460 mg As kg^{-1} dry weight, with concentrations in the reject piles of up to 26,530 mg kg^{-1} (Porter and Peterson 1977).

The availability of arsenic, e.g., the amount of soluble arsenic, is a better indicator of phytotoxicity than total As concentration, since a substantial fraction of this total may not be available and therefore cannot be incorporated (▶ 1, 2, 12, 14). The quantity of soluble or potentially soluble arsenic in soil varies widely with conditions such as pH-value, redox potential, the presence of other elements, namely Fe, Al, and organic matter, and the clay minerals present in the soil (▶ 12). In general, the toxicity effects increase in plants when the soils become more acid, especially with pH values below 5, and As-associated species, such as Fe and Al-oxicompounds, become more soluble.

The As uptake by plants is generally affected by its competition with phosphates in soil or in soil water. Arsenic belongs to the same chemical group as phosphorus and presents a similar geochemical behaviour in soil with the dissociation of its acids and solubility product of its related salts. Low phosphate levels enhance the As desorption from soil particles, thus increasing uptake and phytotoxicity, while large quantities of phosphate in the soil compete with arsenic on the surface of roots, minimizing As uptake and phytotoxicity (Peterson *et al.* 1981). Arsenic phytotoxicity depends on its species. Arsenite, As$^{(III)}$, is generally more phytotoxic than arsenate, As$^{(V)}$ (Burló *et al.* 1999; ▶ 1, 2, 12). Inorganic arsenic inhibits enzymatic activity and As$^{(III)}$ reacts with the sulphhydryl group of proteins, affecting a series of enzymes. Due to the different reactions that involve sulphhydryl groups and phosphorus, As$^{(III)}$ and As$^{(V)}$ may interfere with various physiological and biochemical processes of the plants. Arsenic decreases P-uptake in competition with P, resulting in a reddish leaf colour (Otte and Ernst 1994). Other main features in plants affected by As toxicity are: plant height, number of leaves, number of pods and legume length, inhibited sprout growth (in cereals) and reduced dry mass production.

The toxicity of trace elements in plants appeared in the following decreasing order: As$^{(III)}$ ~ Hg > Cd > Tl > Se$^{(IV)}$ > Te$^{(IV)}$ > Pb > Bi ~ Sb (Ferguson 1990). This sequence, however, depends on soil properties and plant species. Inorganic and organic arsenic from pesticides and herbicides may accumulate in soils and plants. Related sorption occurs mainly through the leaves. Pesticide residues may lead to elevated As levels in soil and phytotoxic effects may remain for long periods of time following their application.

Plants can be classified as accumulating, indicating and excluding specific elements (Brooks 1983). Accumulating plants are capable of absorbing high concentrations of certain elements without any toxic effect. Indicating plants can be used to assess the source and intensity of element concentrations in the soil (Ferguson 1990). The capacity of plants to limit the absorption of a toxic element, irrespective of its quantity in soil, is known as an exclusion mechanism (Brooks 1998). This mechanism can operate at the soil-root interface, within the roots or in the higher parts of the plant. The bioaccumulation capacity of trace-elements varies between species. Plants that accumulate high metal concentrations are referred to as hyperaccumulators. This term was initially used by Brooks *et al.* (1977) to describe plants that absorb and accumulate over 1000 μg of a metal per g, dry mass. Ma *et al.* (2001b) define As-hyperaccumulating plants as those with a bioaccumulation factor (BF) >1, a translocation Factor (TF) >1, and that accumulate above 1 mg As g^{-1} in their biomass. The bioaccumulation factor, defined as the ratio between the element concentration in the biomass (leaves and stem) and in the related soil, is used to determine the effectiveness of the plant in removing metals from soil. The translocation factor, defined as the ratio between the concentration of an element in leaves and in roots, is being used to determine the effectiveness of a plant in transporting the element from the roots to the leaves.

Some fern species have been identified as hyperaccumulators, prompting their study for their potential use in the the recovery of contaminated areas (phytoremediation). Phytoremediation is a general term describing various processes where plants are used to clean or remediate places by removing the polluting agents from soil and water. The plants can break up or degrade organic pollutants and stabilize metallic contaminants acting as filters or traps. Hyperaccumulation of the common fern species *Pteris vittata* was observed first in Central Florida (Komar *et al.* 1998; Ma *et al.* 2001a) at a chromate copper arsenate (CCA)-contaminated location. CCA is widely used in wood conservation. According to the authors, this fern can tolerate As concentrations as high as 1,500 mg g^{-1} in the soil, and As concentrations in the plant may exceed 2.2 wt-% (dry weight), while producing large quantities of biomass (reaching up to 1.7 m height). Besides *Pteris vittata*, various other fern species were identified as As hyperaccumulators, including *Pityrogramma calomelanos* (Francesconi *et al.* 2002), *Pteris cretica* (Ma *et al.* 2001b; Zhao *et al.* 2002), *Pteris longifolia* and *Pteris umbrosa* (Meharg 2003; Zhao *et al.* 2002). According to Francesconi *et al.* (2002), some ferns, including *Pityrogramma calomelanos* are eaten, e.g., in Thailand; the leaves are sautéed or used in salads. The capacity of both fern species *Pityrogramma calomelanos* and *Pteris vittata* to grow healthy in contaminated places and to hyperaccumulate arsenic points to their unprecedented potential for As phytoremediation (Visoottiviseth *et al.* 2002).

Phytoextraction, also called phytoaccumulation, refers to the sorption and transport of metallic contaminants in the soil via the plant roots to its parts above soil (EPA 1998). The selection of plants for phytoextraction should meet the following criteria: resistance to high As concentrations; high bioaccumulation levels, short lifetime, high propagation and biomass levels (Visoottiviseth *et al.* 2002).

Studies by Tu *et al.* (2004) in phytoremediation of As-contaminated groundwater using *Pteris vittata* showed that this fern sorbed arsenic effectively. It was capable of reducing As concentrations to less than 10 μg L^{-1} in three days. The effective implementation of phytoremediation demands an understanding of As-tolerance mechanisms, accumulation and transport in hyperaccumulating plants. Although a lot of research has been conducted into As detoxification and hyperaccumulation in *P. vittata*, it is not clearly understood why this fern species is so effective in As accumulation (Luongo and Ma 2005).

13.2 SAMPLING, SAMPLE PREPARATION AND ANALYSIS

Edible plant material. Backyard cultivation of edible plants is common practice in rural Minas Gerais – as subsistence crops and in larger areas as a source of income. The choice of edible vegetables in this study of As sorption used the following criteria:

- Availability of the vegetables in backyards, which considers low production costs, ease of growth (without fertilizers or agrochemicals) and access to water
- Availability of edible plants during the field work
- The vegetable being part of the population's staple diet
- Time taken to produce the vegetables, distinguishing between rapidly growing (e.g., lettuce), those that can be harvested for up to 12 months (collard green and tubers) and those that are picked late without a specific harvest time.

The samples of edible plants were collected in the Santa Bárbara district (Brumal, Santana do Morro, Barra Feliz, Sumidouro), and in the Nova Lima district (Galo Novo, Matadouro), in the years 2001, 2002, 2004 and 2005, in dry and wet weather periods. Fresh vegetables (cabbage, lettuce, taro root, squash, mustard, parsley, mint, fennel, spinach, tomato, carrot, beetroot, collard-green, sugarcane and sweet potato) were uprooted as a whole and roughly cleaned before transport. The sugarcane for human consumption had been planted at least two years before, allowing the evaluation of As migration in the structure of the vegetable, from the roots to the leaves. As concentrations in soils close to the plant roots were analyzed to study their contribution to the plants.

The plant samples were separated into roots, tubers (if applicable), stem and leaves. A soil sample was also collected at a depth of 20 cm near the roots. When possible, the tubers were dug out with spades after the vegetables were picked manually. The plants were identified and stored in polyethylene bags and taken to the laboratory. Samples were washed in running potable water and rinsed with deionized water to remove possible contaminations such as dust or soil attached to the leaves. The tubers were washed in running water and then scrubbed with a plastic-bristle brush to remove soil particles. The material was then oven-dried at 50°C for 96 h. After drying, the samples were individually homogenized in ceramic-lined mills. All vegetables were oven-dried at 50°C, for 48 h to 96 h to avoid As losses through volatilization. After drying, vegetables and soil samples were ground in ceramic-lined pan mills. The soil samples had been dried in ovens at 50°C to constant weight.

The sample digestion consisted of specific pre-treatment stages needed for the subsequent analysis by flame atomic absorption spectrometry (F-AAS), hydrate generation (HG-AAS), graphite furnace (GF-AAS), and partly inductively-coupled plasma quadrupol mass spectrometry (ICP-QMS). Digestion criteria strictly observed the quality demands (high purity reagents, small volumes, inert materials, and closed digestive systems). The organic plant matter was completely decomposed (Wagner 1994).

The dry plant samples were weighed in triplicates and transferred to flasks, to which 20 mL of concentrated HNO_3 p.a. was added. The digestion was carried out on an electric plate, at a temperature of of 80°C for 4 h, close to complete dryness. This process was repeated with the addition of 10 mL of concentrated HNO_3 p.a. and 5 mL of H_2O_2 p.a. until a clear solution was reached, free of suspension material, and until all organic matter was destroyed. A blank sample was used throughout to control contamination during the procedure, as well as an As reference internal standard sample to check the As recovery and to correct for matrix effects. Soil sub-samples of 1.0 g were combined with 20 mL of concentrated HNO_3 p.a. and 5 mL of HF to dissolve silicates. After dissolution, the samples were dried and 10 mL of concentrated HNO_3 p.a. was added. Parallel to the samples, certified reference material (GXR-6; 330 mg As kg^{-1}) was analyzed. Atomic absorption spectrometry was used to measure As content (Analyst 300, Perkin Elmer). The concentrations given refer to dry mass.

Fern samples. The materials were collected in February and March 2003 in the Mariana districts, townships Monsenhor Horta and Antônio Pereira (Table 13.2), together with related soils. After separating the leaves from the rhizome, the samples were washed in running potable water and rinsed in deionized water. Sprouts, stems and leaves of the fern, *Pteridium aquilinum*, were separated. The sprouts are especially interesting because they are part of the staple diet of the local people.

Lyophilization (freeze-drying) was used to dehydrate the fern samples, a crucial step for their storage. Besides dehydration, lyophilization improves the determination of trace elements in environmental samples. The drying was conducted at −48°C under a vacuum of <133 × 10^{-3} mbar (Model 77520 by LABCONCO). The plant samples were then ground (GlenMills MHM4) and sieved (0.149 mm). The soil samples were dried at room temperature and subsequently sieved. Only the fraction <63 μm (silt and clay) was used in the analysis.

Arsenic was determined by instrumental neutron activation analysis (INAA), applying the k0 method. The samples were irradiated in the research reactor TRIGA MARK I IPR-R1 of CDTN/CNEN in Belo Horizonte. The elements Fe, P, Mn, Rb, and Zn were determined by inductively-coupled plasma quadrupol mass spectroscopy (ICP-QMS; ELAN 5000, Perkin Elmer), following a method described by Daus *et al.* (2005). These elements were determined to evaluate the As bioavailability in the soil matrix.

13.3   ARSENIC IN EDIBLE PLANTS AND FERNS – IRON QUADRANGLE

a) Arsenic determination in lettuce and collard-green. Lettuce, an herbaceous plant (*Asteracea* family, of European and Asian origin) is eaten in salads. It is a source of minerals (calcium and vitamin A) and has been known at least since 500 BC. It is a fast growing vegetable – 45 days between sowing and harvesting. The lettuce samples collected in Santa Bárbara were

Table 13.2. Species of ferns collected, sample locations with UTM coordinates; fern sample and aggregated soil sample identification numbers.

Site ID	UTM coordinates	Fern species and parts analyzed	Fern part ID	Soil ID
MH-01		*Thelypteris* sp. (leaves)	SBC	SM1
		Thelypteris sp. (leaves)	SBT1	SSBT1
TP-01	652 087	*Pteris vittata* (leaves)	SBT2	SSBT1
	7 744 917	*Blechnum* sp. (leaves)	SBT3	SSBT3
TP-09	649 928	*Pteridium aquilinum*	BSC/CSC/FSC	
	7 745 120	(sprout/stem/leaf)		
TP-18	652 101	*Pteris vittata* (leaf/rhizome)	PVTP18/	SPV18
	7 744 961		RPVTP18	
TP-20	651 237	*Pteris vittata* (leaves)	PVTP20	SPV20
	7 745 313			
MH-03	676 762	*Pteris vittata* (leaf/rhizome)	PMH1/RPMH1	SPH1
	7 749 122			
AP-01	658 496	*Pteris vittata* (leaf/rhizome)	PAP1/RPAP1	SPP1
	7 755 080			
AP-02	658 395	*Pteris vittata* (leaf/rhizome)	PAP2/RPAP2	SPP2
	7 755 080			
MC-01	663 945	*Pteris vittata* (leaves)	PMC	SPMC
	7 746 302	*Pityrogramma calomelanos* (leaves)	CMC	SPMC

found to contain less arsenic than samples from Nova Lima; explained by the fact that in Nova Lima parts of the samples had been planted in soil contaminated by old mine tailings. The As contents in lettuce from Santa Bárbara ranged from 0.05 to 6.6 mg kg^{-1}, in close correlation with the As-content in soil (29–467 mg kg^{-1}). In Nova Lima, contents reached 19.9 mg As kg^{-1} in lettuce and in soil up to 14,000 mg As kg^{-1}.

Collard-green, belonging to the family of the *Brassicaceae*, is very rich in nutrients (Vitamins A, C, B1, B2, B5, C, D, E, K; and Fe^{3+}, S, K$^+$, Cl$^-$, Mg^{2+}, Ca^{2+}, and contains a small amount of phosphorus). It came originally from the Mediterranean coast. It is a year-round crop, resistant to plagues. The As concentration in collard greens ranged from 0.05 to 12 mg kg^{-1}, confirming that the highest values occurred in Nova Lima.

b) Arsenic determination in tubers. Being roots, the tubers presented higher As concentrations, due to the proximity of the As-laden soil and the sorbing root surface. The most commonly cultivated tubers in the community studied are the taro root, purple yam and sweet potato. The yam and the sweet potato were picked for sampling only in June 2004, and the taro-root was collected in July 2004 and April 2005. This randomness in the sampling is a result of community-internal dynamics, where each inhabitant chose the different vegetables to be planted.

The sweet potato is the fourth most-consumed vegetable in Brazil. It is a rustic crop, of easy maintenance, resistant to dry spells, easily adaptable, with low production costs and high returns. It belongs to the *Convolvulácea* family from tropical America. It is rich in vitamins A and C and some of the B complex, and it has a great capacity to produce energy per area and time unit (kcal ha^{-1} day^{-1}).

The yam (*Diocoreacea* family) is a rhizome-type root. It may originate from western Africa and was brought to Brazil by the slaves. It is a calorie-rich food, and a source of B vitamins. For this research, only the rhizomes from yam and sweet potato were collected and studied.

The taro-root (*Colocasia esculenta* – *Aracea* family) is a tuber from southeastern Asia. Cultivated since antiquity, it arrived in Brazil in the colonial period via the islands of Cape Verde and São Tomé. The species is adapted to tropical climate and is resistant to the attack of plagues and diseases. It is rich in carbohydrates, complex B vitamins, and mineral salts of

Ca^{2+}, P and Fe^{3+}. It is mainly a small-scale crop, due to the ease of its growth. The taro-roots do not need to be kept under refrigeration because they last for up to 10 weeks in an aerated, dark and dry place. This and the fact that it needs little input make the taro a very popular type of crop. The whole taro-root plant was collected. The results (Table 13.3) show the characteristics of As migration through the entire plant (roots, rubbers, stem, leaves), and confirm that the absorption and translocation mechanism from the roots to the leaves is very effective. These data closely correlate with the As concentrations in the soil. The behaviour is identical with the transport of phosphate through the plants and contributes to the As accumulation in different parts of the vegetable (Tripathi *et al.* 2007).

c) Arsenic determination in sugarcane. This plant has been one of the main crops in Brazil since the times of colonization. It is planted in the Iron Quadrangle for human use (stalk pieces or as juice – *garapa*). Sugarcane, of the *Poaceae* family, is a tall perennial rhizomatous plant that forms clusters. The edible part, the stalk, is characterized by distinct nodes. Plant material from a residential backyard in Santana do Morro was used to evaluate As migration from the soil to the structure of the plant. The plant had been sown two years before collection. The plant was uprooted and separated in eight parts, including roots and leaves. Arsenic concentration in the soil was higher than in the plant, where concentration variances were found along the stalk, between the knots, in the leaves, characterizing As migration dynamics through the plant (Table 13.4). This migration along the structure of sugarcane is similar to that of phosphorus.

d) Arsenic determination in native fern samples. Table 13.5 shows the results for As, Fe, P, Mn, Rb and Zn in the different fern parts and aggregated soil, as well as the translocation factors. "Normal" As concentrations in soils may vary from 0.1–40 mg kg^{-1} (Alloway 1995; ▶ 1, 12). Four from the ten sites had higher concentrations. The highest value at site MC-01 was 966 mg kg^{-1}. This place was a tailings deposit for the gold mine Passagem de Mariana in the past. The soil of site MH-01 (250 mg kg^{-1}) also showed a high As concentration.

Table 13.3. Arsenic concentrations in taro-root plant parts and aggregated soil.

Collected from: Santa Bárbara	As (mg kg^{-1}) dry weight
Leaves	2.57–10.91
Stem	<0.01–4.55
Tuber	<0,01–4.81
Roots	1.42–206.9
Soil adhered to tuber	28.31–246.8
MRC – certified value = 330 mg As kg^{-1}	359

Table 13.4. Arsenic concentration in sugarcane plant sections.

Part of the sugarcane analyzed	As concentration (mg kg^{-1})
Soil adhering to roots	55.7 ± 3.7
Roots	12.2 ± 1.3
1st fraction, close to the roots	<0,05
2nd fraction, central part of cane	8.0 ± 0.11
3rd fraction, central part of cane	12.2 ± 0.09
4th fraction, central part of cane	0.16
5th fraction, central part of cane	0.07
6th fraction, close to the leaves	0.42 ± 0.05
Leaves	1.25 ± 0.09

Sugarcane sample: ca. 2.40 m high. Each fraction: ca. 0.40 m long; 1st fraction starting from root.

Studying the leaves and rhizomes of the *Pteris vittata* samples (TP-18, MH-03, AP-01 and AP-02; Table 13.5), it was found that the elements Fe and Mn had higher concentrations in the rhizome than in the leaves, while the elements As, P, Rb and Zn presented higher concentrations in the leaves. Normally, when soil is the uptake source, metal accumulation in different plant tissues appears in a decreasing order: roots > stem > leaves > fruit > seeds.

High As concentrations were detected in leaves and rhizomes of the fern species *Pteris vittata* and *Pityrogramma calomelanos* (Table 13.5). Concentrations were higher in the leaves than in the rhizomes, with leaf bioaccumulation investigated in *P. calomelanos* only. These results concur with those obtained by Francesconi *et al.* (2002) and Ma *et al.* (2001a) who found that *Pteris vittata* and *Pityrogramma calomelanos* both resist and accumulate arsenic in and from soil. Compared with data in the literature (Cao *et al.* 2004; Francesconi *et al.* 2002; Ma *et al.* 2001b), the BF results in this study present low enrichment values (0.6–27). However, it must be taken into account that there are no data on the age and growth conditions of the native ferns collected. High soil As concentrations do not necessarily correspond to high BF values. This As uptake variability may be connected to the related As availability in the soil. As

Table 13.5. As, Fe, P, Mn, Rb and Zn concentrations ($\mu g\ g^{-1}$ wet weight) in ferns (*Pteris vittata* and *Pityrogramma calomelano*) and related soil ($\mu g\ g^{-1}$, dry weight); bioaccumulation factors (BF) and translocation factors (TF). Data rounded for clarity.

Site	Sample	As			Fe	P	Mn	Rb	Zn
			BF	TF					
MH-01	SBC (leaf)	0.8			650		42		24
	SM1 (soil)	250			243,600		5,360		64
	SBT1 (leaf)	0.1			214		8.6		11
	SSBT1 (soil)	27			190,100		2,930		70
	SBT2 (leaf)	17	0.6		310		25		13
TP-01	SSBT1 (solo)	27			190,100		2,930		70
	SBT3 (leaf)	<0.1			144		19		8
	SSBT3 (soil)	20			26,000		250		33
	BSC (sprout)	<0.1			5		2	3.6	5
TP-09	CSC (stem)	<0.1			6		5	3.9	4
	FSC (leaf)	<0.1			38		5	10	15
	PVTP18 (leaf)	130	2.6	1.7	461	540	54	67	109
TP-18	RPVTP18 (riz.)	77			2,740	130	75	12	59
	SPV18 (soil)	49			222,000	1,070	12,014	78	209
TP-20	PVTP20 (leaf)	140	7.6		106	330	37	43	43
	SPV20 (soil)	18			264,000	410	2,940	65	46
	PMH1 (leaf)	90	7.0	3.6	169	750	29	56	63
MH-03	RPMH1 (riz.)	25			183	240	6	9	36
	SPH1 (soil)	13			53,000	560	340	25	70
	PAP1 (leaf)	330	27	2.4	1,850	590	58	51	52
AP-01	RPAP1 (riz.)	140			4,650	190	120	13	74
	SPP1 (soil)	12			294,000	330	1,140	106	30
	PAP2 (leaf)	180	1.8	2.4	5,420	770	290	75	52
AP-02	RPAP2 (riz.)	76			7,780	560	450	25	43
	SPP2 (soil)	100			457,000	730	6750	10	38
	PMC (leaf)	2,300	2.4		200	440	75	107	83
MC-01	SPMC (soil)	970			326,000	520	3,750	24	46
	CMC (leaf)	1,740	1.8		320	690	47	141	55

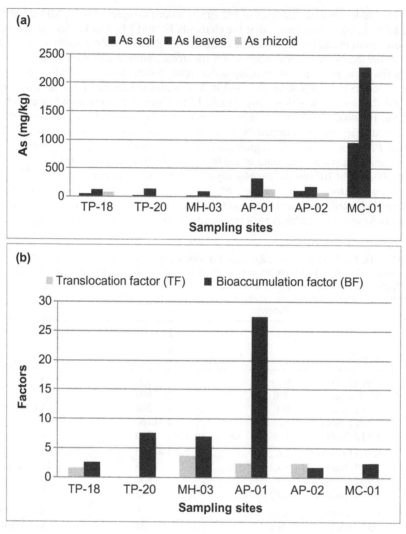

Figure 13.1.   (a) Arsenic concentrations in leaves, rhizomes and soil and (b) bioaccumulator factors (BF) and translocation factors (TF) for As in *Pteris vittata* samples from different sites.

shown, the highest BF values were found at the sites TP-20, MH-03 and AP-01, where Mn and Fe-concentrations were lower as compared with other sites (Figure 13.1 and Table 13.5).

Kabata-Pendias and Pendias (2000) reported that soil Fe-oxides have a strong negative impact on As availability (▶ 12), explained by the strong As sorption onto Fe-oxihydroxides. Deschamps *et al.* (2002) demonstrated that the Mn and Fe-enriched soil of the Iron Quadrangle present significant sorption capacity for the trivalent and pentavalent As oxidation states, thus potentially contributing to the reduction of dissolved arsenic in water and consequently its bioavailability (▶ 3, 4, 16, 17).

The presence of phosphorus in soils can compete with the As sorption by plants. The highest BF value (27.4) occurred at site AP-01, with lower P-concentrations. Therefore, the low phosphate content in the soil may lead to a relatively high As sorption by the plants. Rubidium (BF 0.5–7.5) and Zn (BF 0.5–0.8) also accumulated in some *Pteris vittata* samples. Rubidium, a rarely studied alkaline metal, can be an essential ultratrace element for humans and other organisms. Zinc is one of the seven micronutrient elements (Zn, Cu, Mn, Fe, B, Mo and Cl), essential for normal plant nutrition (Campbell *et al.* 2005). The evaluation of the As, Fe, P, Mn, Rb and Zn results in the ferns species *Thelypteris sp.*, *Blechnum sp.* and *Pteridium aquilinum* (Table 13.5) did not find any significant acccumulation of either element. In *Pteridium aquilinum* only, the Fe, Rb and Zn-concentrations were higher in the leaves (FSC).

13.4   CONSEQUENCES OF ARSENIC IN EDIBLE PLANTS

The As enrichment in the edible parts of plants differed widely between the plant species. Values for lettuce, collard-green and tubers were above 1 mg kg^{-1}, reaching 10.9 mg kg^{-1} (Table 13.6). This agrees with elevated values found in the Iron Quadrangle by Birmann (2002), Haßler (2002), Hilscher (2003) and Raßbach (2005). While higher values are more representative for the Nova Lima district (<0.1–20 mg kg^{-1}) than the Santa Bárbara district (<0.1–7 mg kg^{-1}), individual high values were also found in vegetable gardens in Brumal, Santa Bárbara district (7 mg kg^{-1} in lettuce).

A calculation of potential human exposure enables establishing an association with some possible health effects (▶ 9, 14). This is helpful in designing health programmes for an exposed population (▶ 15). To estimate a true threat to human health, based on the ingestion of plants, is not an easy task and should not be used exclusively. In exposed populations, many more aspects need to be considered (▶ 14). When a substance is carcinogenic, however, there are no safe limits of exposure, and it is important to bear in mind that an entire population may potentially be at risk. The World Health Organization (WHO) adopted a reference PTWI value (Provisional Tolerable Weekly Intake). This weekly intake of 15 μg As kg^{-1} body weight is the maximum tolerable value, above which harmful effects cannot be excluded. To calculate the maximum tolerable intake, a differentiation between children and adult (average) body weight is needed – and again may lead to substantial bias, e.g., when considering malnourished or otherwise stressed segments of a population. With body weights of 70 kg (adult), and 20 kg (child), these figures lead to a PTWI of 1.050 μg As (adult), and for a child of 300 μg As. With an actual realistic weekly ingestion of 200 g of vegetables analyzed in this study, and the encountered As concentrations, an average exposure can be given (Fig. 13.2).

Based on these results, Figure 13.2 and Table 13.6 demonstrate that the tolerable limit for children is already being exceeded with the consumption of lettuce alone. Based on that

Table 13.6.   Mean As concentration values in collard-green, lettuce and soil, and PTWI percentage based on the weekly consumption per person.

Plant	Sampling site	As (mg kg^{-1}) (dry weight)	WI (μg week^{-1} person^{-1})	PTWI (%) Child	PTWI (%) adult	Soil As (mg kg^{-1})
Collard-green	Brumal	0.18	13	4	1	83
	Barra Feliz	0.53	37	12	4	42
	Santana do Morro	0.07	5	2	<1	37
	Sumidouro	0.05	3	1	<1	34
	Galo Novo	2.38	166	55	16	360
	Matadouro	3.86	270	90	16	1,600
Lettuce	Brumal	1.62	113	38	<1	83
	Santana do Morro	0.1	5	2	2	37
	Galo Novo	4.46	312	104	30	360
	Matadouro	10.9	761	254	72	1,600

PTWI: Provisional Tolerable Weekly Intake; WI: Weekly Intake (▶ 14).

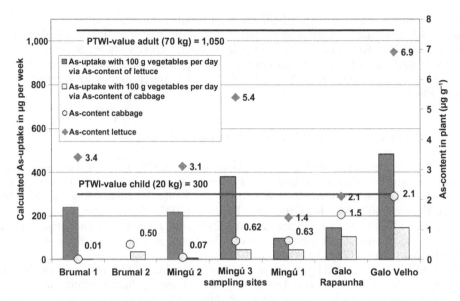

Figure 13.2.   Weekly mean As ingestion by children (20 kg body weight) and adults (70 kg body weight), based on conservative amounts of collard-green (cabbage) and lettuce consumption.

scenario, the presence of lettuce as a staple in their diet, children reach 250% of their PTWI in Nova Lima district, and adults 72% of their PTWI. One should keep in mind that this food-based scenario is but one pathway for potential As intake (▶ 10, 11, 12). However, the individual and detected body As loads can be explained by this ingestion source (▶ 14). In consequence, the Project recommended to local authorities that all private vegetable gardens on As contaminated soils in Galo Novo be eliminated and replaced by elevated gardens (▶ 17).

REFERENCES

Alloway, B.J. (ed): *Heavy metals in soils*. 2nd ed. Blackie Academic & Professional, London, 1995, p.390.

Birmann, K.: *Gesamtarsenbestimmung in ausgewählten Gemüsepflanzen*. B.Sc. Thesis, Unpublished, TU Bergakademie Freiberg, Freiberg, Germany, 2002, p.59 + annex.

Brooks, R.R.: *Biological methods of prospecting for minerals*. Wiley, New York, NY, USA, 1983, p.322.

Brooks, R.R.: *Plants that hyperaccumulate trace metals*. Wallingford Cab International, UK, 1998, p.380.

Brooks, R.R., Lee, J., Reeves, R.D. & Jaffre, T.: Detection of nickeliferous rocks by analysis of herbarium specimens of indicator plants. *J. Geochem. Explor*. 7 (1977), pp.49–57.

Burló, F., Guijarro, l., Carbonell-Barrachina, A.A., Valero, D. & Martinez-Sanchez, F.: Arsenic species: effects on and accumulation by tomato plants. *J. Agric. Food Chem.* 47:3 (1999), pp.1247–1253.

Campbell, L.M., Fisk, A.T., Wang, X., Köck, G. & Muir, D.C.G.: Evidence for biomagnification of rubidium in freshwater and marine food webs. *Can. J. Fish Aquat. Sci.* 62:5 (2005), pp.1161–1167.

Cao, X., Ma, L.Q. & Tu, C.: Antioxidative responses to arsenic in the arsenic-hyperaccumulator Chinese brake fern (*Pteris vittata* L.). *Environ. Pollut.* 128:3 (2004), pp.317–325.

Daus, B., Wennrich, R., Morgenstern, P., Holger, W., Palmieri, H.E.L., Nalini, H.A. Jr., Leonel, L.V., Monteiro, R.P.G. & Moreira, R.M.: Arsenic speciation in plant samples from the Iron Quadrangle, Minas Gerais, Brazil. *Microchim. Acta* 151:3–4 (2005), pp.175–180.

Deschamps, E., Ciminelli, V.S.T., Lange, F.T., Matschullat, J., Raue, B. & Schmidt, H.: Soil and sediment geochemistry of the Iron Quadrangle, Brazil. The case of arsenic. *J. Soils Sediments* 2:4 (2002), pp.216–222.

Dudka, S. & Miller, W.P.: Accumulation of potentially toxic elements in plants and their transfer to human food chain. *J. Environ. Sci. Health B* 34:4 (1999), pp.681–708.

EPA: A citizen's guide to phytoremediation. United States Environmental Protection Agency 542-F-98-011, 1998.

Fiedler, H.J. & Rösler, H.J. (eds): *Spurenelemente in der Umwelt*. VEB Gustav Fischer, Jena, 1987, p.278.

Francesconi, K., Visoottiviseth, P., Sridokchan, W. & Goessler, W.: Arsenic species in a hyperaccumulating fern, *Pityrogramma calomelanos*: a potential phytoremediator of arsenic-contaminated soils. *Sci. Total Environ.* 284:1–3 (2002), pp.27–35.

Haßler, S.: *Geochemie von Böden und deren Wechselwirkung mit Pflanzen im südlichen Eisernen Vierecks, Brasilien*. Masters thesis, Unpublished, TU Bergakademie Freiberg, Freiberg, Germany, 2002, p.124 + annex.

Helgesen, H. & Larsen, E.H.: Bioavailability and speciation of arsenic in carrots grown in contaminated soil. *Analyst* 123 (1998), pp.791–796.

Hilscher, A.: *Biogeochemie von Wild- und Nahrungspflanzen von As-belasteten Standorten in Minas Gerais und Sachsen*. M.Sc. thesis, Unpublished, TU Bergakademie Freiberg, Freiberg, Germany, 2003, p.135.

Kabata-Pendias, A. & Pendias, H.: *Trace elements in soils and plants*. 3rd ed. CRC Press, USA, 2000, p.432.

Komar, K.M., Ma, L.Q., Rockwood, D. & Syed, A.: Identification of arsenic tolerant and hyperaccumulating plants from arsenic contaminated soils in Florida, 1998. Agronomy Abstract, p.343.

Larsen, E.H., Moseholm, L. & Nielsen, M.M.: Atmospheric deposition of trace elements around point sources and humam health risk assessment: II. Uptake of arsenic and chromium by vegetables grown near a wood preservation factory. *Sci. Total Environ.* 126:3 (1992), pp.263–275.

Luongo, T. & Ma, L.Q.: Characteristics of arsenic accumulation by Pteris and non-Pteris ferns. *Plant and Soil* 277:1–2 (2005), pp.117–126.

Ma, L.Q., Komar, K.M., Tu, C., Zhang, W., Cai, Y. & Kennelly, E.D.: A fern that hyperaccumulates arsenic. *Nature* 409:6820 (2001a), p.579.

Ma, L.Q., Komar, K.M. & Kennelly, E.D.: Methods for removing pollutants from contaminated soil materials with a fern plant, 2001b. USA Patent US patent No. 6 302 942. Date issued: 10/16/01.

Matschullat, J.: Arsenic in the geosphere – a review. *Sci. Total Environ.* 249:13 (2000), pp.297–312.

Meharg, A.A.: Variation in arsenic accumulation hyperaccumulation in ferns and their allies. *New Phytol.* 157:1 (2003), pp.25–31.

O'Neill, P.: Arsenic. In: Alloway, B.J. (ed): *Heavy Metals in Soils*. Blackie Academic & Professional, London, 1995, pp.105–121.

Otte, M.L. & Ernst, W.H.O.: Arsenic in vegetation of wetlands. In: Nriagu, J.O. (ed): *Arsenic in the Environment. I. Cycling and Characterization*. Wiley, New York, 1994, pp.365–379.

Porter, E.K. & Peterson, P.J.: Arsenic tolerance in grasses growing on mine waste. *Environ. Pollut.* 14:4 (1977), pp.255–265.

Pyles, R.A. & Woolson, E.A.: Quantitation and characterization of arsenic compounds in vegetables grown in arsenic treated soil. *J. Agric. Food Chem.* 30:5 (1982), pp.866–870.

Raßbach, K.: *Arsentransfer und Human-Biomonitoring in Minas Gerais, Brasilien*. M.Sc. thesis, Unpublished, TU Bergakademie Freiberg, Freiberg, Germany, 2005, p.93 + annex.

Tripathi, R.D., Srivastava1, S., Mishra, S., Singh, N., Tuli, R., Gupta, D.K. & Maathuis, F.J.M.: Arsenic hazards: strategies for tolerance and remediation by plants. *Trends Biotechnol.* 25:4 (2007), pp.158–165.

Tu, S., Ma, L.Q., Fayiga, A.O. & Zillioux, E.: Phytoremediation of arsenic-contaminated groundwater by the hyperaccumulating fern *Pteris vittata* L. *Internat. J. Phytoremed.* 6:1 (2004), pp.35–47.

Visoottiviseth, P., Francesconi, K., Sridokchan, W.: The potential of Thai indigenous plant species for the phytoremediation of arsenic contaminated land. *Environ. Pollut.* 118:3 (2002), pp.453–461.

Wagner, G.: Biologische Umweltproben. In: Stoeppler, M. (ed): *Probenahme und Aufschluss*. Springer Labormanual, 1994, pp.71–83.

Zhao, F.J., Dunham, S.J. & McGrath, S.P.: Arsenic hyperaccumulation by different fern species. *New Phytol.* 156:1 (2002), pp.27–31.

CHAPTER 14

Human biomonitoring

Nilton Couto, Silvânia Mattos & Jörg Matschullat

14.1 INTRODUCTION

Exposure to toxic agents may lead to their uptake in certain amounts by biota including humans (inhalation, ingestion and skin permeation). Toxins may be distributed in physiological tracts and tissues, metabolized and/or excreted fully or partially (Barr *et al.* 2005; ▶ 2, 12, 13). To assess the extent of an exposure and related health risks, biological indicators, such as cellular alterations, biochemical or molecular biological fluids can be used (Vine 1996; ▶ 12.3).

Three types of biological indicators are used to assess exposure, effect and susceptibility. However, there are no clear distinctions between them. As a rule, biological media like blood and urine, alveolar air and tissue (hair and fingernails) or individual cells (e.g., skin or inner organs) may be used to measure a toxin concentration in potentially exposed individuals (Diabaté 2008; Vine 1994), and to study their reaction. Ideally, indicators would be applied that are capable of providing a reliable link between the concentration of a toxin in a specific compartment of the organism and a related health effect. It would be even better to have an indicator that helps to anticipate an undesirable physiological reaction or a disease, and that could differentiate between the toxic agent alone and related consequences from nutritional and physiological factors (sex and age), lifestyle (nutrition, hygiene, etc.), behaviour, and differences in individual susceptibility. This would be most helpful since these characteristics may lead to significant differences in individual resilience and physiological reaction. Unfortunately, no such indicator exists today, and all bio-assays and sophisticated cell tests are not yet capable of delivering such simple answers. Yet, biological indicators are still the best option to evaluate the health of a given population. The toxicokinetic properties of a toxin need to be known before selecting an appropriate indicator (▶ 2). To give a simple example, blood tests to assess arsenic are possible, but not very reliable, whereas urine tests present the method of choice.

Some studies take into account characteristics related to lifestyle, especially focused on nutritional habits. A study by Yamauchi *et al.* (1992) in Japan shows a normal mean As ingestion value of $195 \pm 235 \, \mu g \, day^{-1}$. The range was between 16 and 1,040 μg As day^{-1}, distributed to 83% organic As species versus 17% inorganic species – typical for seafood-based diets. Similar data were found in Catalonia, Spain, where As uptake reached as much as 224 μg As day^{-1} (Llobet *et al.* 2003). Data from Canada and people whose staple diets consisted of fish and shellfish (64%), mostly from fresh water, presented an average of 38 μg As per person and day (Dabeka *et al.* 1993). Based on such data, Anke (1993) defined the normal oral As ingestion rate for an adult as 10 to 170 μg As day^{-1}, depending on the quantity of fish and seafood. The mean As concentration in the human body is very low ($<100 \, \mu g$ As g^{-1}; Table 14.1; ▶ 1). With more intense exposure, these values can increase to 2 mg As kg^{-1} (Anke 1993; Iffland 1994).

Studies of As levels in blood are restricted to assess very recent acute exposure at higher levels or direct poisoning. Its trigger – effect correlation is too weak to monitor low exposure levels for a longer period of time. The As levels in blood have a half-life of 60 hours and are generally below 1 $\mu g \, L^{-1}$. Levels as high as 1,000 $\mu g \, L^{-1}$ have been registered and even higher levels in cases of death (ATSDR 2004). The rather high value range given by Bowen (1979) is certainly an artefact and due to limited analytical capability at the time (Table 14.1).

Arsenic may accumulate in hair and fingernails or toenails, as it is associated with the cysteine sulphhydryl in keratin (▶ 2). Organic As compounds, originating from marine animals, do not accumulate in these tissues. Therefore, respective As concentrations reflect an exposure to inorganic arsenic only. Nevertheless, hair and fingernails/toenails may be used to evaluate exposure retrospectively (as far back as one year), although erroneous results may easily occur through external sample contamination. Non-contaminated values are generally below 1 mg L^{-1}, but concentrations may increase by a hundred times and remain high for 12 months after the exposure. Non-exposed individuals were found to have a mean concentration of 0.06 μg As g^{-1} in their hair (Hirner *et al.* 2000). This value may increase with exposure. Up to 12.4 μg As g^{-1} hair were detected in people using contaminated groundwater in China (Schmitt *et al.* 2002). As a whole, values departing from 1.2 μg g^{-1} represent enhanced As exposure (Anke 1986), while WHO (2001) considers amounts between 0.4–0.8 μg g^{-1} as normal and sets a minimal critical limit of 1 μg g^{-1}. If organic As species are not incorporated to keratin, hair reflects only exposure to inorganic arsenic (Vahter 1998).

On the other hand, arsenic absorbed through airways and the gastrointestinal tract is mainly eliminated through kidney action and the urinary tract (Vahter 1999). Elimination of arsenic and its methylated metabolites occurs mainly through urine, glomerular filtration, tubular secretion and active reabsorption (Ford 1994), with little contribution of the gall. Excretion of arsenic starts two to eight hours after intake of a single As dose, and can last ten days. This period of time may extend to 70 days when As input is repeated (Larini *et al.* 1997). The elimination speed also depends on the As species, how it enters the body, and on the dose. Pentavalent inorganic compounds taken orally are rapidly eliminated. Slow excretion increases the As load in the body. A deficiency of the glutation enzyme that is part of the methylation process reduces the As elimination level. Organo-arsenical compounds are essentially not metabolized and are eliminated without any substantial alteration. Arsenobetaine (AsB) and tetramethylarsenic (TMAs) dominate in many foods and do not experience biotransformation in human beings, being almost completely eliminated via urine.

Urinary arsenic "As$_U$" refers to the sum total of the common inorganic As species. Their concentration usually remains below 10 μg As L^{-1} urine. Regardless of extent and type of exposure, As concentrations in urine consist to 10–30% of inorganic As species, 10–20% of monomethylarsonic acid (MMA), and 60–80% of dimethylarsinic acid (DMA; Vahter 1999). To assess recent As exposure, As$_U$ concentrations are the most reliable indicator to evaluate a human population. This indicator also helps identifying residents in the proximity of significant As sources. As$_U$ also correlates well with As concentrations in a workplace environment. Sampling is relatively easy (▶ 14.1) and poses neither ethical nor biomedical issues. It is important to take into account the presence of non-toxic organic As forms in urine that

Table 14.1.   Mean As concentrations in human tissues (mg kg^{-1}) and biological fluids (μg L^{-1}).

Tissue	As concentrations (mg kg^{-1})	
	1–3*	4, 5*
Kidneys	≤100*[1]	0.005–1.5
Liver	≤100*[1]	0.02–1.6
Muscular tissues	≤100*[1]	0.009–0.65
Bones	<100*[1]	0.08–1.6
Hair	0.01–0.6*[1]	0.02–3.7
Finger and toenails	0.01–0.6*[1]	0.2–3
Blood	<0.5–4.0 μg L^{-1}*[2]	1.7–10 μg L^{-1}
Urine	6.4 μg L^{-1}*[3]	< 1–8 μg L^{-1}

*[1]Iffland (1994), [2]Yamauchi *et al.* (1992), [3]Becker *et al.* (2002) estimated arithmetic mean for the German population with n = 4,741, range from [4]Bowen (1979) and [5]Fergusson (1990).

may influence the results. The excretion of arsenobetaine, for example, increases after the ingestion of some seafood. This must be considered when selecting the appropriate analytical method for the sample. The analysis of As species, however, is by no means simple and requires highly qualified personel and a complex laboratory infrastructure.

The reference value (RV), adopted by Brazilian Law for As concentration in urine, is 10 µg As g^{-1} of creatinine; the maximum allowed biological index (IBMP) equals 50 µg As g^{-1} of creatinine for recent exposures (Brasil 1994; ▶ 5). The adjustment of the As concentration using creatinine aims at correcting the effects of the dilution by the daily variation observed in the individuals. Various authors (Becker *et al.* 2001; Krause 1984; Krause *et al.* 1987, 1996; LGA 1997; Matschullat 2000; Matschullat *et al.* 2000; Seifert *et al.* 2000; and UBA 2003) adopted an As concentration classification for human urine based on the following toxicological categories (Table 14.2).

The main effects of chronic As exposure are keratosis and skin hyperpigmentation. A certain peripheral neuropathy, or reduction of the speed of nerve conduction amplitude, was also found by some authors (ATSDR 2004; ▶ 2). It is difficult to detect this effect in exposed populations and without clear clinical signs characteristic of toxicity. Some enzymes that act in the synthesis and degradation of the heme may alter when exposed to arsenic. There are signs of an increase in the levels of uroporphyrin, coproporphyrin and bilirubin, which could be used as effect indicators. However, these increases are not specific, since other toxic metals may cause similar effects.

The development of an As biomonitoring project for the districts of Nova Lima and Santa Bárbara, Minas Gerais, Brazil, was initially prompted by the knowledge of potentially high As concentrations related to the gold mineralizations (▶ 7) and by individual voices in related communities that expressed concern about possible related As exposure and health effects (▶ 8). The idea to set up a related study, the ARSENEX project, was successfully submitted to the Ethics and Research Committee of the State Health Agency "Fundação Ezequiel Dias" (FUNED; ▶ 8), since biological samples had to be used to evaluate the target population.

The study was conducted between 1998 and 2005, and concentrated on a selection of six townships in the Nova Lima district (Galo Novo, Galo Velho, Matadouro, Mina d'Água, Mingú and Resende) and five townships in the Santa Bárbara district (Barra Feliz, Brumal, Sumidouro, Carrapato and Santana do Morro). Whenever possible, the urine samples were collected twice a year, alternating between the regional rainy and dry seasons. The campaigns were mainly carried out in public schools, with authorization of the municipal and state secretaries for health and education, and the substantial support of local teachers, authorities and the resident population (▶ 15).

14.2   SAMPLING, STORAGE, AND ANALYTICAL PROCEDURES

Preparation. The field sampling team received preliminary training, and jointly developed a sampling strategy, taking the specific situation of each community into account. To effectively cover a representative part of the population and to avoid bias, the focus was on children between 8 and 12 years of age. Children run higher risks when exposed to environmental toxins since their metabolism works faster and their defense mechanisms are not yet fully

Table 14.2.   Toxicological categories, based on the total As concentrations in human urine.

Category	Comments	As (µg L^{-1})
I	Inconspicuous	<15
II	High; no immediate threat, monitoring recommended	15–40
III	Very high; health hazard, demands detailed investigation	>40

Krause *et al.* (1987).

developed (this is further ponounced in children up to 6 years of age, who show much more frequent hand-mouth contact). The children in the selected age group had not yet reached puberty and were very open to participate in the study. Children at this age do not generally consume alcohol, nicotine, or other drugs, which helps to exclude uncertainties or heterogeneous data (e.g., with subcritical groups of society from other age groups). Another advantage of selecting this age group was the collaboration of their schools and the willingness of the teachers to assist in the effort. Older and younger people, usually relatives, parents and neighbours of the children, also volunteered and participated in the study.

To prepare the population for the campaign, headmasters and teachers were contacted, and meetings were arranged with authorities and parents. The purpose of the investigation was explained and discussed at these meetings to obtain support and participation. Volunteers and legal representatives of the children were informed of the need to sign an authorizing document on the day of sampling, to permit the use of the urine samples and the resulting data. This initial contact was crucial and guaranteed lasting support for the project. A questionnaire to further support the subsequent evaluation and interpretation of the results was prepared, and distributed to parents and children (Figure 14.1). Basic information such as date of birth, weight, sex, and data on nutrition, health, and lifestyle (which can vary considerably from place to place) as well as on possible chronical or acute diseases was gathered. Many children represented the first generation in their families with unrestricted access to formal education. These children played the role of interpreters for their parents, which lent a wider social and psychological dimension to the project, which had not been foreseen (▶ 9, 15).

Sampling, storage and transport. Generally, all steps followed the IUPAC standard and related guidelines (Cornelis *et al.* 1995). The heads of the schools organized the sampling, with entire class groups in sequence. The children handed in their questionnaires, were registered, and received a urine flask. Each questionnaire, jointly with the urine flask, received a code (pre-prepared sticker). Upon return from the lavatories, each child was thanked for her or his contribution, was given a small present and some questions were posed to confirm the replies on their individual questionnaire. This communication proved very useful, not only in respect to verify information but mainly to add an additional human dimension to a process that otherwise would have been too fast and impersonal, particularly for younger children.

The sampling team consisted of at least three persons. One collected the questionnaires, assigned the code numbers, filled in the participant list and handed out the urine flasks. A second person handled the samples, and a third person verified questionnaire answers and talked to the children. In total, more than 1,600 urine samples were collected for the project (Table 14.3), which is a representative number for the age groups and considerably higher than in many similar studies. Each urine flask was visually checked for volume (minimum ca. 20 mL) and optical characteristics (colour, sediments, etc.). Related observations were added

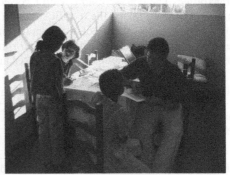

Figure 14.1.   Questionnaire verification with school children prior to sampling (Photos Michelle Amorim).

as a note to the participant list. The sample was then split: an aliquot of 2 mL was set aside for the creatinine tests, and the remainder was filled after acidification (for sample conservation) into previously acid-washed, screw-cap, HDPE tubes (20 mL).

Samples with creatinine values below 30 mg dL^{-1} and above 250 mg dL^{-1} were discarded, due to excessive dilution or unusual concentration. Within minutes (after cooling to room temperature), both bottles were stored at approximately 4°C. At the end of each day, the samples were taken to a freezer at FUNED, where they remained frozen (−20°C) until analysis, or further transport and handling. Before another transport, the frozen samples were placed in a thermally-insulated box and all spaces between samples were filled with dry ice. The samples remained frozen solid over more than 24-hours, when they reached other laboratories in Stuttgart (State Health Agency Baden-Württemberg) and Freiberg (TU Bergakademie Freiberg), both in Germany. Transport regulations for human urine are relatively liberal (Thurm *et al.* 2007). At Freiberg, samples were stored at a temperature of −18 to −20°C until analysis. All samples were thawed to room temperature (approximately 20°C) to adapt to the ambient environment prior to analysis. Unused sample aliquots were refrozen (−18 to −20°C) and stored for future replication analysis or additional work. To better compare the results, samples of a control group (28 adults, who did not live in the region and had no known As exposure) were included in the sample collective.

Hair samples were collected less consistently and only from 99 children. A small amount of hair from the occipital region was cut with stainless steel scissors, carefully avoiding upper hair. The lower back of the head is more protected under the hair cover, and the removal of a sample is almost imperceptible. The hair samples were stored in small Whirlpack® plastic bags without any further treatment. Hair colour and type were documented. The samples were collected during August 2003: 32 in the Santa Bárbara district and 67 in the Nova Lima district. Sixty percent of the samples were from women or girls.

Analysis and quality control. The creatinine analysis was done immediately upon sample arrival in the laboratory, using Bioclin/Labtest kits with UV/VIS spectometry (HP 8451A, Hewlett Packard, USA). The As analyses at FUNED were processed within a week by atomic absorption spectrometry (AAnalyst 300), with a flow injection system (FIAS 400), an automatic sampler (AS90), and an EDL system 2 electrode-less discharge lamp – all Perkin Elmer). In Stuttgart, atomic absorption spectrometry with hydride generation (4100 with MHS-20, Perkin Elmer) was used, and in Freiberg inductively-coupled quadrupol mass spectrometry (ICP-QMS, Perkin Elmer) was used. The obtained total As values represent the sum of the toxicologically important species, As$^{(III)}$, As$^{(V)}$, and methylated As acids (MMA and DMA). Generally, the species were not differentiated. All As values were above the quantification limit of 0.7 µg L^{-1}. Individual species analyses were performed at FUNED after the implementation of a methodology by Guo *et al.* (1997), capable of quantifying toxic As forms in urine (As$^{(III)}$, As$^{(V)}$, MMA, and DMA). More detailed species analyses were done in selected samples with colleagues in Essen, Germany (Rabieh *et al.* 2008).

All reagents were at least of "pro analysis" quality, except for the hydrochloric acid, which was metal-free. All solutions were prepared with purified water (Milli-Q system). The calibration curves were between 1.0 and 20.0 µg As L^{-1}, based on an As standard stock solution of 1,000 µg mL^{-1} (Merck, Germany). Samples were processed at FUNED with the addition of 2 mL of L-cysteine 5% and 7 mL of hydrochloric acid 0.03 mol L^{-1} to 1 mL of urine. After resting for one hour to allow for a complete reaction, the samples were analyzed. A solution of 0.03 mol L^{-1} hydrochloric acid was used to carry the compounds formed, and a mixture of a solution of sodium hydroxide 0.5% and sodium borohydride 0.5% was used as a shrinking solution (Guo *et al.* 1997). Instrument readings were obtained at a wavelength of 193.7 nm, a lamp current of 380 mA, with a quartz cell at 900°C, and a carrier gas flow of 40 to 50 mL minute^{-1} (Perkin Elmer, 1996). Lower detection (3σ) and quantification limits (10σ) of the methodology were calculated using the standard deviations (LLD: 0.2 µg L^{-1}, LLQ: 0.7 µg L^{-1}).

The certified reference materials NIST (CRM 2670) and BIO Lyphochek (Urine Metals Control, level 1: 50 µg As L^{-1}, and Level 2: 154 µg As L^{-1}) were used jointly as unknown samples with reagent blanks for analytical quality control (Matschullat *et al.* 2000). CRM was analyzed with each batch. Whenever available, samples were submitted to triplicate analysis. The As-related results from the analyses of the certified samples – NIST and Lyphochek – were within the minimal and maximum limits established, indicating adequacy of the analytical methodology used (Figures 14.2 and 14.3).

Additionally, FUNED participated in a comparative study between two laboratories, using 18 urine samples with unknown As content, the same methodology, but different equipment, reagents and analysts (Fig. 14.4). The excellent correlation between the independent results (r^2: 0.99015) confirmed the analytical methodology.

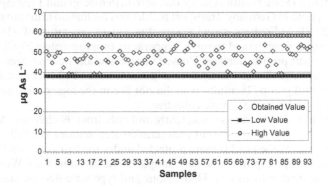

Figure 14.2.   Results of the NIST-certified samples.

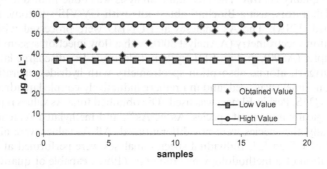

Figure 14.3.   Results of the Lyphochek-certified sample.

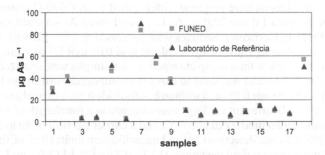

Figure 14.4.   Results of the interlaboratory studies (Laboratório de Referência = reference laboratory).

All hair samples were analyzed in Freiberg. After washing for 30 min. in distilled water, samples were air-dried at constant air humidity (60% RH), and subsequently pulverized (ultra centrifugal mill ZM 1000, Fritsch, Germany). After lyophilization (Alpha-I, Christ, Germany), the samples were digested in ultrapure nitric acid in a microwave oven (temperature was steadily increased to 210°C, 15 min.; MLS-Ethos, Germany); up to 10 mL of ultrapure water were added. As concentrations in the hair samples were determined by atomic absorption spectrometry with a graphite furnace (GF-AAS) and ICP-QMS (Perkin Elmer Elan 9000, Canada). The certified reference materials BCR 397 (Europe) and GBR 601 and 9101 (China) for hair samples were used for quality control. Recovery rates were between 105 and 110% in every series (Junghänel 2003).

14.3 RESULTS AND INTERPRETATION

Urine samples. All following data are based on creatinine-corrected values. The correlation between total As values (μg As L^{-1}) and creatinine-corrected values (μg As g^{-1} creatinine) ranged from 0.699 to 0.798 (p = 0.05). The adult sub-collective will not be discussed in detail since their number remained insufficient. Most adult participants were women, since most men were working away from home at collection time and because most women were more cooperative. The following data and discussion is exclusively related to the children in the investigation.

When looking at the entire data set (Table 14.3), the average As concentrations are heterogeneously distributed, reflecting the already known structural differences between the two districts (Table 14.4). At the same time, interannual variation between the mean As values in urine can be observed, along with a more substantial drop between the first campaign in 1998 and all subsequent years (Table 14.5).

When differentiating further between townships, significant differences emerged between the localities (Table 14.6). The same is true for the individual sub-collectives from each township, where differences of over four orders of magnitude may occur (not shown). Results of the control group were included in Table 14.6 with 2.0–8.1 μg As g^{-1} creatinine, and a mean value 3.5 μg As g^{-1} creatinine.

Table 14.3. Sampling frequency.

Month-Year	Location	Number of samples	Responsible institution
04–1998	NL, SB	126	TUBAF
08–1999	NL, SB	273	TUBAF
08–2001	NL, SB	294	TUBAF
12–2001	SB	29/29	TUBAF/FUNED
04–2002	SB	156/127	TUBAF/FUNED
12–2002	SB	216/199	TUBAF/FUNED
08–2003	NL, SB	292/140	TUBAF/FUNED
12–2003	SB	101	FUNED
06–2004	SB	36	FUNED
05–2005	SDM	53	FUNED
06–2005	SDM	29	FUNED
08–2005	SDM	29	FUNED
Total		1,634	

NL: Nova Lima; SB: Santa Bárbara; SDM: Santana do Morro; FUNED: Fundação Ezequiel Dias; TUBAF: TU Bergakademie Freiberg, Germany.

Table 14.4.   Mean As concentrations and ranges (µg L⁻¹) in human urine by district.

Nova Lima	Santa Bárbara	Total Population
13.6	7.7	10.3
(1.0–58.0)	(0.2–30.6)	(0.2–58.0)

Table 14.5.   Interannual variances mean As values in human urine from 1998, and 2003–2005.

Year	Mean value As (µg g⁻¹ creat.)	Maximum Value (µg g⁻¹ creat.)
1998	25.7	106
2003	9.01	34.8
2004	11.1	99.1
2005	11.8	64.3

Table 14.6.   Mean As values in urine (µg As g⁻¹ creatinine), per township and district.

	Santa Bárbara					Nova Lima		
	Barra Feliz	Brumal	Sumidouro	Carrapato	Santana do Morro	Galo	Matadouro	Control
Mean	14.6	14.2	14.6	8.7	11.2	31.4	12.0	3.5
75% perc.	18.5	16.8	24.7	12.0	14.7	35.7	11.8	4.7
n	190	282	30	36	57	58	17	28

n = Number of samples ("Galo" includes Galo Novo and Galo Velho).

Values below As_U occurred at every location, clearly demonstrating that even on the local level, a generalization is impossible. The mean values, however, demonstrate that the samples from the Nova Lima district yield considerably higher concentrations than those from the Santa Bárbara district. The high As values in both districts indicate individual risks, related to As dissipation in the environment and to lifestyle. Figure 14.5 presents the percentiles of individuals with As contents below the reference values (<RV), above the Maximum Allowed Biological Value (MABV) and between these two limits (>RV <MABV) in both regions.

About half (47%) of the individuals from the Santa Bárbara district presented As_U-results below the Reference Value (RV), and only 1% exceeded the MABV. In Nova Lima, however, exposure was evident in most of the population, while 16% were above the maximum index allowed. Figure 14.6 displays the same data, differentiated by age group. Here, adults are included in Nova Lima, despite the statistically insufficient group size. The highest As levels were found in Galo township (Nova Lima), where values above MABV at every age level indicate As exposure and contamination (mean value 31.4 µg As g⁻¹ creatinine, maximum 126.4 µg As g⁻¹ creatinine). The lowest values occured in the Carrapato township (Santa Bárbara; Table 14.6). Yet in the townships Barra Feliz, Brumal and in Santana do Morro (Santa Bárbara), most adolescents and children showed As values above RV. Apart from that, the values exceeded MABV in 0.7% of the samples collected from the children in Barra Feliz, in 4.5% of the samples of the adolescents from Brumal and in 7.7% Santana do Morro. These data suggest a potential As source in these communities (Fig. 14.6). Table 14.2 presents the risk classes that were adapted in this project. Table 14.7 and Figure 14.7 display the data by these risk categories (class I to III).

The population of Nova Lima is obviously more As exposed than that of the Santa Bárbara district. The lowest mean As concentrations in urine occurred in Matadouro, which

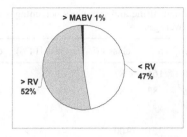 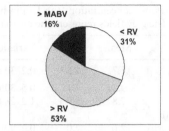

Figure 14.5. As_U results. Left: Santa Bárbara district. Right: Nova Lima district. See text.

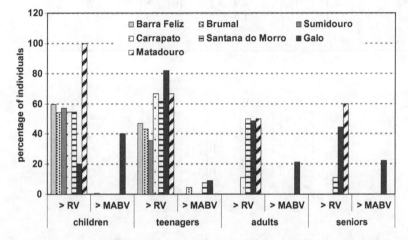

Figure 14.6. Percentile of individuals with As values above RV, per age range ("Galo" represents Galo Novo and Galo Velho).

also showed the narrowest range and a low median value. This township seems directly comparable to the average conditions in the Santa Bárbara district. The roads are paved, and in general, the socioeconomic conditions appear more urban than rural (▶ 9). Samples from the Galo townships (Nova Lima) yielded substantially higher concentrations, where 1/5 to 1/3 of the population fall into risk class II. Both Galo Novo and Galo Velho are located close to the site of a mining operation. Part of the population of Galo lives on one of the largest old mine tailing deposits, close to Morro do Galo. The residues of this deposit contain As concentrations in weight-% (▶ 11), and an As trioxide factory (closed in 1975) was active for decades in Morro do Galo. The tailings deposit was rehabilitated in 2003 and the lands, gardens and yards of the houses were covered with concrete, successfully reducing further dissipation of As contaminated soil and mine tailing materials (▶ 17). In every sampling campaign since 1998, the subcollective of Galo Novo showed the highest As rates in human urine. The percentile distribution (Fig. 14.8) gives a good idea of the As exposure levels in the different regions. The control group (4.7 µg As g⁻¹ creatinine) is clearly separated from all other sub-collectives (11.8 to 35.7 µg As g⁻¹ creatinine).

The townships Sumidouro and Santana do Morro (Santa Bárbara district) represent by far the lowest average As concentrations, exclusively within class I (Table 14.7). The two townships lie to the south of the Conceição river and next to the Caraça nature reserve, where dust distribution is insignificant and the historical small-scale mining seems irrelevant (▶ 7). The district average is considerably higher with 10 to 20% in class II. The highest As levels

Table 14.7.   Arsenic concentrations (μg L^{-1}) in human urine and risk class percentages (sampling in August 2003). The grey fields represent the district averages.

Locality	Mean	Median	Variation	n	class I (%)	class II (%)	class III (%)
Santa Bárbara	7.7	5.5	0.2–30.6	165	87	13	–
Brumal	9.1	7.1	0.5–30.6	83	81	19	–
Barra Feliz	7.8	5.9	1.2–28.0	57	91	9	–
Sumidouro with Santana do Morro	2.6	2.0	0.2–9.9	21	100	–	–
Nova Lima	13.6	11.2	1.0–58.0	131	68	29	3
Galo Novo	15.6	12.3	1.6–58.0	96	61	35	4
Galo Velho	16.1	10.5	4.9–24.2	10	80	20	–
Matadouro	7.1	6.7	1.0–22.6	25	92	8	–
Total (n = 292)	10.3	7.9	0.2–580	292	79	20	1

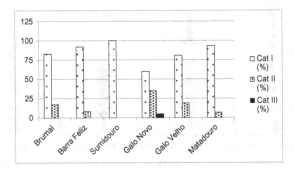

Figure 14.7.   Risk class distribution by township, data from August 2003.

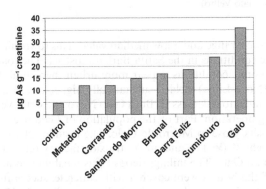

Figure 14.8.   75th percentiles of all townships. "Galo" stands for both G. Novo and G. Velho.

occurred in Brumal, followed by Barra Feliz, which may well relate to their location near historical or active gold mining sites.

It must be noted, however, that most of the people in these townships drank water from springs without treatment (▶ 11). These waters present high levels of coliform bacteria, an indicator for fecal contamination. The arsenic occurs mainly in the solid phase due to the prevailing oxidizing conditions. Its presence in suspended solids during the rainy season deserves special attention. This, along with the high load of coliform bacteria were the main reasons

Table 14.8. Total As concentration and representative distribution of the different species in urine of selected children, collected in August 2003 (Rabieh et al. 2008).

As concentrations (μg As L⁻¹)

ID	As(V)	As(III)	MMAs(V)	MMAs(III)	DMAs(V)	AsB	Sum	Total	Total	Rec.
1	2.03±0.21	2.27±0.34	3.49±0.48	nd	6.85±0.89	0.64±0.06	15.3±1.8	17.7±0.3	17.3	86.3
2	2.92±0.43	4.86±0.61	4.38±0.43	nd	12.95±1.13	7.99±0.26	33.1±1.8	38.0±0.5	33.8	87.1
3	4.69±0.61	5.28±0.41	16.23±1.97	2.00±0.16	20.45±1.66	nd	48.7±4.3	55.2±0.5	58	88.1
4	1.56±0.24	2.31±0.33	1.47±0.26	0.42±0.06	9.16±0.65	nd	14.9±1.3	16.7±0.5	17.8	89.3
5	2.73±0.40	3.79±0.41	2.27±0.17	0.58±0.07	8.37±0.71	nd	17.7±1.7	21.3±0.2	20	83.3
6	1.48±0.26	3.64±0.39	5.71±0.61	0.89±0.08	5.28±0.46	0.69±0.07	17.7±1.5	19.2±0.4	22.6	92.1
7	3.06±0.55	5.81±0.42	11.69±0.92	0.74±0.08	6.57±0.56	0.71±0.02	28.6±1.6	30.9±0.4	32.6	92.5
8	2.15±0.39	3.12±0.22	3.25±0.29	0.62±0.06	4.38±0.66	nd	13.5±1.5	16.1±0.3	15.2	84
9	1.49±0.21	4.36±0.35	5.01±0.61	0.69±0.06	5.72±0.51	nd	17.3±1.7	18.2±0.3	22.9	94.9
10	2.95±0.42	6.71±0.69	7.36±0.40	1.18±0.13	14.97±0.82	0.76±0.07	33.9±2.4	38.5±0.3	41	88.1
11	2.57±0.36	4.75±0.31	nd	0.59±0.08	8.12±0.49	nd	16.0±0.9	19.3±0.2	17.3	83.1
12	2.45±0.46	5.17±0.46	7.27±0.41	0.97±0.06	8.59±0.46	0.47±0.03	24.5±1.3	31.4±0.2	27	79.4
13	1.69±0.28	nd	2.56±0.29	nd	10.38±0.87	nd	14.6±1.3	18.9±0.1	18	77.4
14	2.26±0.23	4.19±0.34	6.58±0.70	0.74±0.05	7.35±0.41	0.94±0.06	22.1±1.4	26.3±0.2	24.2	83.9
15	3.51±0.36	3.00±0.39	2.86±0.36	0.67±0.07	14.92±0.76	nd	25.0±1.7	27.2±0.6	30.7	91.8

nd = not detected; [a]The values presented are the means of four replicates each; [b]Sum of the concentration of the individual species; [c]Total concentration obtained by ICP-MS (n = 10); [d]Total concentration obtained by ICP-QMS at TUBAF (Freiberg/Germany), [e]Rec. = recovery (%) was calculated as: Sum$_{species}$ × 100 / Total concentration[c]; from Rabieh et al. (2008).

to construct a water treatment plant in Santana do Morro – a non-foreseen result of this project, which brought many benefits (▶ 16).

To further evaluate the toxicological characteristics of the As enrichment in the urine samples, a species analysis was carried out at Essen University, Germany (Rabieh *et al.* 2008). The test determined nine inorganic species: penta- and trivalent arsenic, monomethylarsinic acid (MMAs$^{(V)}$), monomethylarsonic acid MMAs$^{(III)}$), dimethylarsinic acid (DMAs$^{(V)}$), dimethylarsonic acid (DMAs$^{(III)}$), arsenobetaine (AsB), arsenocoline (AsC), and trimethylamine oxide (TMAO). While AsC, DMAs$^{(III)}$ and TMAO were not detected in any sample, all other species mostly occurred, with 11.6% a 45.8% (mean of 29.4%) of the inorganic species (Table 14.8).

Supposing that the relative distribution of As species in the urine was equal to the corporal distribution (around 30% of the total of the species found appeared in the most toxic inorganic form, in the ratio of As$^{(III)}$ to As$^{(V)}$ of 0.85–2.93; mean 1.72).

Hair samples. Hair samples illustrate the difference between the areas under study in the two municipalities. While in Santa Bárbara only one person (71 years old) was found to have high As concentrations in the hair, 23% of 26 people in Nova Lima had high concentrations as a result of an atypical As exposure. Even so, the unequal gender and age variation are not enough to produce a conclusive correlation with the risk evaluation (Table 14.9).

14.4   HEALTH ISSUES AND CONCLUSIONS

The obtained analytical results show that people in the communities of the Santa Bárbara and Nova Lima districts were exposed to arsenic, albeit with very large individual differences. Children and adolescents appear to be more at risk than adults. The situation was very severe in one township only, the neighborhood of Galo (Nova Lima). Here, As values in urine were above the threshold in every age group. The township is close to a derelict As trioxide factory in Morro do Galo that was closed down in 1975. Part of the residential area sits on one of the oldest mine tailings deposits (▶ 11.1). The results of the hair samples corroborate the results of the urine analysis.

However, it is important to point out that health symptoms (observed in parallel questionnaires + interviews) in the two groups, differ particularly between Brumal (Santa Bárbara) and Galo (Nova Lima). In the former, the most common symptoms were associated with the respiratory system, and in the latter, there is a higher incidence of muscular and skin problems (Fig. 14.9; ▶ 9), suggesting that the most relevant As source in each of these areas may be different. The main As sources in Brumal most likely relate to contaminated soils (▶ 12). The situation in Galo appears more complex, since As exposure is common in every age group and environmental compartment. The additional exposure associated with the old mine tailings – before the mitigation took effect (▶ 17), may play a crucial role in individual exposure and related As uptake.

Several chronic diseases were mentioned in the questionnaires, both in the human biomonitoring and the socioeconomic study (▶ 9). Around 6% of the individuals in Brumal reported a notable incidence of keratosis, associated or not with headaches and/or bronchitis, and skin discolouration. However, no direct relationship with the As values in their urine could be observed (mean 15.6, range 3.2 to 36.7 μg As g^{-1} creatinine). In the neighborhood of Galo, however, only four individuals complained of skin stains, three of whom had high As contents in their urine, while the others were asymptomatic. Fifty-one individuals complained about some or several of the symptoms nausea, diarrhea, muscular pains, abdominal pains, weakness, headaches or loss of sensitivity of the hands and feet. Even in individuals with As$_U$ below recommended values (RV), disorders such as headaches and abdominal pains or loss of appetite were frequent complaints. One individual with 80.5 μg As g^{-1} creatinine had white streaks in the fingernails. Symptoms related to the respiratory system were not very frequent

Table 14.9.   Mean As concentrations and ranges (μg g^{-1}) in human hair.

	Total population	Nova Lima	Santa Bárbara
Mean	1.1	1.2	0.6
Ranges	(<0.5–6.6)	(<0.5–6.6)	(<0.5–0.6)

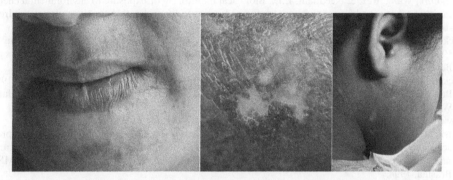

Figure 14.9.   Dark (left) and light (center and right) skin discolourations of unknown origin on faces of a middle-aged woman (left) and a young girl (right). Similar discolourations appear on arms, chest (center) and legs (photos Jörg Matschullat).

in the group with As values above RV. Ten individuals in Galo and one in Matadouro complained about hypertension. The As levels in their urine were above RV, varying from 10.3 to 123.5 μg As g^{-1} creatinine. In the neighborhood of Galo, As$_U$ was found to be above the threshold in 25% of this group.

It is important to point out that among the sampled individuals with normal As values in urine, there are asymptomatic cases with results above RV and MIA, and cases in which the symptoms are not specific. It is important to remember that the risk indicator used in the ARSENEX project is more adequate for group evaluations, to investigate recent exposure in a certain region or community. The consumption of fish in the studied regions is insignificant and had no correlation with the data obtained (questionnaire-based and ▶ 9). A slight increase was observed in values obtained in the dry season as compared to those from the rainy season. This result was anticipated, as the tendency for elevated As concentration in environmental compartments increases in the dry season (dust, water concentration, etc.).

As for the difference between men and women, results were rather differentiated. The mean As values in the urine of men (9.4 μg As g^{-1} creatinine) and women (8.8 μg As g^{-1} creatinine) in Carrapato near Barra Feliz showed no significant difference. In Sumidouro (Santa Bárbara: ♂ = 18.1 μg As g^{-1} creatinine; ♀ = 5.0 μg As g^{-1} creatinine), and in Galo (Nova Lima: ♂ = 35.4 μg g^{-1} creatinine; ♀ = 25.6 μg As g^{-1} creatinine) distinct differences emerged and the As values in men were higher than in women.

Even though not all questions surrounding the As-related risks could be answered, this human biomonitoring study proved its worth in various ways. This study drew the attention of local inhabitants and made them very much aware of the potential risks from both the natural presence of arsenic and enhanced dissipation due to mining. Local and regional authorities also obtained a better understanding of the related risks of contamination and the potential health hazards. This encouraged consideration of concrete measures to curb or minimize exposure and led to both changes in behaviour (▶ 15) and to technical solutions for remediation and mitigation (▶ 16, 17).

REFERENCES

Anke, M.: Arsenic. In: Mertz, W. (ed): *Trace elements in human and animal nutrition*, Volume 2. Academic Press, New York, 1986, pp.347–372.

Anke, M.: Die Bedeutung der Spurenelemente für die Fauna. In: Fiedler, H.G. & Rösler, H.J. (eds): *Spurenelemente in der Umwelt*. 2nd ed. Gustav Fischer, Jena, 1993, p.385.

ATSDR: Toxicological profile for Arsenic. Draft, Agency for Toxic Substances and Disease Registry, U.S. Department of Health and Human Services, 2004. www.atsdr.cdc.gov/ToxProfiles/.

Barr, D.B., Wang, R.Y. & Needham, L.L.: Biological monitoring of exposure to environmental chemicals throughout the life stages: requirements and issues for consideration for the national children's study. *Environ. Health Perspect.* 113:8 (2005), pp.1083–1091.

Becker, K., Kaus, S., Helm, D., Krause, C., Meyer, E., Schulz, C. & Seiwert, M.: Umwelt-Survey 1998, Band IV: Trinkwasser. Elementgehalte in Stagnationsproben des häuslichen Trinkwassers der Bevölkerung in Deutschland. WaBoLu-Heft 02/2001, Umweltbundesamt, Berlin, 2001.

Becker, K., Kaus, S., Krause, C., Lepom, P., Schulz, C., Seiwert, M. & Seifert, B.: Umwelt-Survey 1998, Band III: Human-Biomonitoring. Stoffgehalte in Blut und Urin der Bevölkerung in Deutschland. WaBoLu-Heft 01/2002, Umweltbundesamt, Berlin, 2002.

Bowen, H.J.M.: *Environmental chemistry of the elements*. Academic Press, London, 1979, p.348.

Brasil: Secretaria de Segurança e Saúde no Trabalho. Portaria n° 24, de 29 de dezembro de 1994. Aprovação da Norma Regulamentadora n° 7 que estabelece os parâmetros mínimos e diretrizes gerais a serem observados na execução do Programa de Controle Médico de Saúde Ocupacional – PCMSO. Brasília, 1994.

Cornelis, R., Heinzow, B., Herber, R.F.M., Molin Christensen, J., Paulsen, O.M., Sabbioni, E., Templeton, D.M., Thomassen, Y., Vahter, M. & Vesterberg, O.: Sample collection guidelines for trace elements in blood and urine. *Pure Appl. Chem.* 67:8–9 (1995), pp.1575–1608.

Dabeka, R.W., McKenzie, A.D., Lacroix, G.M.A., Cleroux, C., Bowe, S., Graham, R.A., Conacher, B.S. & Verdier, P.C.: Survey of arsenic in total diet food composites and estimation of the dietary intake of arsenic by Canadian adults and children. *J. AOAC. Int.* 76:1 (1993), pp.14–25.

Diabaté, S., Mülhopt, S., Paur, H.R. & Krug, H.F.: The response of a co-culture lung model to fine and ultrafine particles of incinerator fly ash at the air-liquid interface. *Altern. Lab. Anim.* 36:3 (2008), pp.285–298.

Fergusson, J.E.: *The heavy elements: chemistry, environmental impact and health effects*. Pergamon Press, Oxford, 1990, p.614.

Ford, M.: Fast automated determination of toxicologically relevant arsenic in urine by flow injection-hydride generation atomic absorption spectrometry. In: *Toxicologic Emergencies*. 5th ed.: Appleton & Lange, Norwalk, 1994, p.72.

Guo, T., Baasner, J. & Tsalev, D.L.: Fast automated determination of toxicologically relevant arsenic in urine by flow injection-hydride generation atomic absorption spectrometry. *Analyt. Chim. Acta* 349 (1997), pp.313–318.

Hirner, A.V., Rehage, H. & Sulkowski, M.: *Umweltgeochemie*. Steinkopff Verlag, Darmstadt, 2000, p.836.

Iffland, R.: Arsenic. In: Seiler, H.G., Sigel, A. & Sigel, H. (eds): *Handbook on clinical and analytical chemistry*. Marcel Dekker, New York, 1994, pp.237–253.

Junghänel, I.: *Untersuchung von Humanhaar zur Beurteilung fer Arsenproblematik im Eisernen Viereck, Brasilien*. M.Sc. thesis, Unpublished TU Bergakademie Freiberg, 2003, p.107.

Krause, C.: Untersuchungen zur Arsenexposition verschiedener Personengruppen. *Schriftenreihe WaBoLu* 59 (1984), pp.201–214.

Krause, C., Thron, H.L., Wagner, H.M., Flesch-Janys, D. & Schümann, M.: Ergebnisse aus Feldstudien über die Belastung der Bevölkerung mit Schwermetallen durch industrielle Quellen. *WaBoLu-Hefte* Volume 74. Fischer Verlag, Stuttgart, 1987, pp.105–111.

Krause, C., Babisch, W., Becker, K., Bernigau, W., Hoffmann, K., Nölke, P., Schulz, C., Schwabe, R., Seiwert, M. & Thefeld, W.: Umwelt-Survey 1990/92, Band la: Studienbeschreibung und Human-Biomonitoring: Deskription der Spurenelementgehalte in Blut und Urin der Bevölkerung der Bundesrepublik Deutschland. WaBoLu-Hefte 1/1996.

Larini, L., Salgado, P.E.T. & Lepera, J.S.: Metais. In: Larini, L. (ed): *Toxicologia*. 3rd ed. Manole, São Paulo, 1997, pp.131–135.

LGA: Pilotprojekt Beobachtungsgesundheitsämter 1992/93–1994/95. Zusammenfassender Bericht über die dreijährige Pilotphase. Landesgesundheitsamt Baden-Württemberg im Auftrag des Sozialministeriums Baden-Württemberg, Stuttgart, 1997.

Llobet, J.M., Falcó, G., Casas, C., Teixidó, A. & Domingo, J.L.: Concentrations of arsenic, cadmium, mercury and lead in common foods and estimated daily intake by children, adolescent adults, and seniors of Catalonia, Spain. *J. Agric. Food Chem.* 51:3 (2003), pp.838–842.

Matschullat, J.: Arsenic in the geosphere – a review. *Sci. Total Environ.* 249 (2000), pp.297–312.

Matschullat, J., Borba, R.P., Deschamps, E., Figueiredo, B.R., Gabrio, T. & Schwenk, M.: Human and environmental contamination in the Iron Quadrangle, Brazil. *Appl. Geochem.* 15:2 (2000), pp.181–190.

Perkin Elmer: Flow injection mercury/hydride analyses. Recommended analytical conditions and general information. Release 4.0, Germany, 1996.

Rabieh, S., Hirner, V. & Matschullat, J.: Determination of arsenic species in human urine using high-performance liquid-chromatography (HPLC) coupled with inductively-coupled plasma-mass spectrometry (ICP-MS). *J. Anal. Atomic. Spectrometry* 23:3 (2008), pp.544–549.

Schmitt, M.T., Mumford, J.S., Xia, Y., Ma, H., Ning, Z. & Le, C.: Health effects of chronic exposure to arsenic via drinking water in inner Mongolia. *Abstract 5th Internat Conf arsenic exposure and health effects*, 14–18 July San Diego, 2002.

Seifert, B., Becker, K., Helm, D., Krause, C., Schulz, C. & Seiwert, M.: The German Environmental Survey 1990/1992 (GerES II): reference concentrations of selected environmental pollutants in blood, urine, hair, house, dust, drinking water and indoor air. *J. Exposure Anal. Environ. Epidemiol.* 10:6 (2000), pp.552–565.

Thurm, V., Schoeller, A., Mauff, G., Just, H.M. & Tschäpe, H.: Versand von medizinischem Untersuchungsmaterial. Neue Bestimmungen ab 2007. *Deutsches Ärzteblatt* 104:46 (2007), pp.A1–A7.

UBA: Stoffmonographie Arsen – Referenzwerte für Urin. Stellungnahme der Kommission "Human-Biomonitoring" des UBA. Bundesgesundheitsblatt – Gesundheitsforschung – Gesundheitsschutz, Volume 46. Springer, Verlag, 2003, pp.1098–1106.

Vahter, M.: Arsenic. In: Clarkson, T.W., Friberg, L., Nordber, G.F. & Sager, P.R. (eds): *Biological monitoring of toxic metals*. Plenum Press, New York, 1998, pp.303–321.

Vahter, M.: Variation in human metabolism of arsenic. In: Chappell, W.R., Abernathy, C.O. & Calderon, R.I. (eds): *Arsenic exposure and health effects*, Volume III. Elsevier, Amsterdam, 1999, pp.267–279.

Vine, M.F.: Biological markers: their use in quantitative assessments. *Adv. Dental Res.* 8:1 (1994), pp.92–99.

Vine, M.F.: Biologic markers of exposure: current status and future research needs. *Toxicol. Industrial Health* 12:2 (1996), pp.189–200.

WHO: Arsenic. Environmental health perspectives, Geneva, 2001, p.224.

Yamauchi, H., Takahashi, K., Mashiko, M., Saitoh, J. & Yamamura, Y.: Intake of different chemical species of dietary arsenic by the Japanese, and their blood and urinary arsenic levels. *Appl. Organo-metallic Chem.* 6 (1992), pp.383–388.

Section IV
Solutions and outlook – meeting the challenges

CHAPTER 15

Environmental and health education

Sandra Oberdá, Eleonora Deschamps & Leonardo Fittipaldi

Education for sustainable development intends to broaden the scope of environmental education by taking into account economic and social issues, in order to consider environmental affairs within the framework of an efficient economy and a just society (sustainability triangle). This approach was adopted by the government of the Federal Republic of Germany in 2002 (National Sustainability Strategy). At a global level, the United Nations announced the Decade of Education for Sustainable Development (DESD) 2005–2014.

http://www.bildungsserver.de/zeigen_e.html?seite = 5592

15.1 ENVIRONMENTAL EDUCATION

Throughout its existence, humankind has been using the natural resources of planet Earth in a rather insensible way, not considering that its resources are finite. Environmental (and health) problems have emerged, particularly following the Industrial Revolution of the 18th Century, resulting from the pollution of air, water, and soils for an ever-increasing population. In the last decades, especially as of the 1970s, environmental degradation and deterioration of the quality of life have been the cause of serious concern. Key publications spread awareness and led to a growing and more critical distance towards "eternal material growth", e.g., Carson (1962), Meadows and Meadows (1972), Council on Environmental Quality (1980). At the same time, many groups, including NGOs and national and international institutions, started to address the need to develop environmental education.

Environmental education is an eye-opener, helping to reverse, or at least curb human ignorance towards the environment by providing incentive to gaining both more knowledge about environmental issues and supporting a more sustainable way of life. As compared to merely science or "fact"-based approaches, it should be noted that emotional or non-rational approaches to a deeper understanding may often lead to more success and awareness. Depending on perspective, commitment to environmental education is often related to a "new civilizing project" that promotes ethics surrounding life. This demands reflection and action to combat inequality and poverty and to promote access to common goods and services (clean water, healthy food, medical care and electricity). Encouraging society to act, based on the principles of justice, equality, democracy and sustainability, is one of the main aims of this type of environmental education. A more grass-roots approach to environmental education brings people closer to reality and leads them to an understanding of complex inter-relationships between the human being and their biophysical and cultural universe (Obara et al. 2005). According to the Brazilian law 9.795, article 5 (Brasil 1999), which enforces the national policy of environmental education, the main aims of which are to:

- Develop an integrated understanding of the environment in its multiple and complex relationships, involving ecological, psychological, legal, political, social, economic scientific, cultural and ethical aspects
- Democratize information about the environment
- Strengthen a critical approach to environmental and social issues

- Give incentive to permanent and responsible individual and collective participation in the preservation of environmental balance, while at the same time understanding the defense of environmental quality as part of the exercise of citizenship
- Stimulate the cooperation between different national regions, at micro and macro levels to build an environmentally-balanced society, based on the principles of freedom, equality, solidarity, democracy, social justice, responsibility and sustainability
- Encourage and strengthen the integration between science and technology
- Strengthen citizenship, self determination of the people, and solidarity as fundaments for the future of humanity.

Three types of environmental education are generally described in national and international literature: a) formal environmental education at school; b) non-formal education, including environmental aspects; and c) informal education comprising information and communication, characterized by outdoor activities, involving flexibility of methods and content and a very diversified target population with regard to age, schooling, knowledge of environmental issues, etc. (Porto 1996). Obviously, other manifestations exist that may contribute to public awareness of environmental issues.

Environmental education has become an important tool in changing people's behaviour, and has proven effective in fostering meaningful development. To be effective, however, an environmental education programme should promote attitudes and skills needed for the preservation and improvement of environmental quality, in addition to the development of mere knowledge. The challenge is to build a basis for the understanding of the complex nature of the environment, and to encourage respect for all life-forms. "*Environmental education should reach everyone – inside or outside of schools; in the community, in religious, cultural, sports and professional associations – it must go where the people are*" (Dias 2004).

Key to raising awareness are educational workshops (*oficina* in Portuguese) that cater individually to specific communities and target groups (children, adults, multiplicators). *Oficina* (from Latin *officina*) means a shop, a factory; a crafts room. Symbolically, it may refer to a place or session where people meet to develop their skills and abilities under the guidance of qualified professionals, using didactic-pedagogic materials and equipment for the teaching-learning process. "*... the oficina, as a didactic tool, is a space for reflection and discussion of experience, making it possible for educators and learners to overcome the obstacles and barriers in the pedagogic practice, with prospects for a new transforming praxis*" (Obara *et al.* 2005).

15.2   ENVIRONMENTAL EDUCATION AND HEALTHCARE WORKSHOPS

To contribute to the improvement of life quality, educational actions were developed for the communities and participating schools within the ARSENEX project. The purpose was to raise the participating groups' awareness of the most important issues related to local As contamination. This included advice on its prevention. Important problems affecting the community as a whole were always discussed. Oliveira (1989) comments that "*... knowledge alone is not enough. The basis for environmental education lies in the levels of commitment and participation*". Each workshop put substantial emphasis on motivating the participants to feel responsible as multiplicators and ambassadors of environmental understanding. Environmental education workshops, lectures, courses and the distribution of educational booklets contributed to the ARSENEX project's aims.

It was decided to join the workshop sessions with the urine sampling campaign so that the team could contact the parents of the public school students, and the villagers. The idea was to encourage permission from the parents to collect urine from their children and for the team to explain As-related issues at the same time. The As issues were always set into a broader context, from local to global, and contextualized with local characteristics on geology, mining activities, contamination types and risk reduction. The participants, parents, teachers, headmistresses, headmasters, and other citizens were encouraged to express their doubts and ask questions in open discussions. Both formal and informal environmental education workshops were offered.

The formal activities were developed with students of the public schools of Brumal, Barra Feliz and Santa Bárbara (Santa Bárbara district), and in Mingú and Galo (Nova Lima district), with substantial support and active participation of the school principals. These school heads made the workshop venues available and mobilized students, teachers and parents by encouraging them to participate in the activities. They also helped by spreading the word about the workshops in the townships. The non-formal education workshops were held in community association rooms and targeted the local community. People of all ages participated in the workshop, as well as government representatives, community leaders and professionals in the fields of environment, sanitation and health. The community and health agents played a very important role in developing the non-formal educational events.

The educational activities were designed to meet the needs identified by the environmental perception study (▶ 9) and the expectations of the public school educators. Another goal was to meet the demands of local authorities, community agents and healthcare professionals. Twenty-three environmental education workshops were developed, which addressed eight topics related to As contamination:

1. Arsenic-related information, including contaminant pathways
2. Importance of personal hygiene
3. Regional As contamination and the installation of a water treatment plant
4. Cleaning water reservoirs at home
5. Education on water usage
6. Preventing As contamination through environmental dust
7. Garbage – its impact on environment, health, and a source of income
8. Waste incineration and the greenhouse effect.

Every workshop addressed the As theme, as it was the main focus of the Project. Twenty-one workshops were held in Santa Bárbara and two in Nova Lima, each session lasting about one hour (Oberdá *et al.* 2007: Table 6.1).

15.3   WORKSHOP DETAILS

The Project coordinator opened each workshop with a project overview that informed and also motivated the participant's engagement. Then the participants were given information about the work being developed in the community: its origin, the area covered, the environmental and biological compartments studied (human health), As-related issues, and urine collection (▶ 14). The activity-of-the-day was then introduced, along with its importance and aims. Questions raised by the participants were answered in a simple, clear and above all honest and open manner. The innovative character of the ARSENEX project and the acquisition of knowledge with the active participation of the people involved were explained. At the same time, the team worked to motivate the participants to take an active part in this knowledge build-up. The methodological path of the project was explained as well as its aims. The following description of the individual workshops presents the respective aims and describes the means to make them a reality.

1.   Arsenic-related information, including contaminant pathways
This workshop began with a general view of As-related issues in the world and in Brazil. Local characteristics were included and pollution pathways illustrated. Information on the dangers posed by arsenic were complemented with concrete advice on As-related individual risks, types of exposure and opportunities for prevention. Frequent questions and doubts regarding the urine sampling were addressed and discussed. During the workshops, participants also shared information, such as details on the location of springs, allowing the team to better design and adapt its sampling scheme.

Workshops were used to deliver more detailed information to the local medical community and the Secretary of Health of Santa Bárbara. Here, the latest research on human toxicology and As intoxication were presented and discussed. The professionals were motivated to support the study and to contribute with their experience, such as epidemiological observations. Research results from the ARSENEX project were presented and the methods that FUNED and the project team applied were explained and discussed. Selected urine analysis results were discussed and the medical community was asked to assist in distributing and explaining these results to the people. The Health Secretary of Santa Bárbara expressed great interest in the work and immediately offered to cooperate by appointing healthcare professionals to join the project. Figure 15.1 gives an impression from these meetings.

2. Importance of personal hygiene

The topic of personal hygiene in the local communities was defined to meet demands of a health agent from Brumal, who identified the need for such educational work. Awareness on the importance of good personal hygiene (bathing, hair, face, hand, feet care, sweat, etc.), including care and cleanliness in the home, were presented as ways to prevent diseases and reduce risks of As exposure. This included orientation on male and female intimate hygiene and a discussion of problems arising from poor hygienic habits, and also orientation on habits in general, care with foods, private and public places and pets.

The participants were advised on how to clean their houses, to prevent dust from building up and instructed about precautions to be taken with children when playing on unpaved ground. The importance of participants acting as multipliers of the newly acquired knowledge

Figure 15.1.   Lecture for health professionals at the Municipal Health Secretary of Santa Bárbara (photos Michelly Amorim).

Figure 15.2.   Workshop for community and health agents, and for Brumal citizens at the Residents Association of Brumal (photos Michelly Amorim).

was highlighted, and they were asked to pass all information on to their students, members of the community and family. Figures 15.2 and 15.3 show photos of workshop activities.

3. Regional As contamination and the installation of a water treatment plant

The topic was based on the apparent need to raise people's awareness in Santana do Morro of the importance of using treated water to combat persistent problems (waterborne diseases; ▶ 14). The workshop informed the citizens about project results, and particularly about the planned construction of a water treatment plant in their township (▶ 16). The program included

Figure 15.3. Community workshop at the Residents Association of Santana do Morro (photos Michelly Amorim).

Figure 15.4. Left: urine collection. Right: signature of consent for urine collection (▶ 14) in Galo (photos Michelly Amorim).

Figure 15.5. Left: sugarcane (*Saccharum* spec.) collection. Right: Taro roots (*Colocasia esculenta*) collection for As analysis (▶ 13) in Santana do Morro (photos Michelly Amorim).

a description of the project activities in the district such as, collection of water, soil, fruit and vegetables (Fig. 15.5) as well as urine and hair (Fig. 15.4). The reason for installing a treatment plant was explained in detail to motivate the participants to take responsibility for the good performance of the plant. At the end of the session, a slideshow with pictures taken in Santana do Morro (by project members) increased further receptiveness and acceptance for the project.

4. Cleaning water reservoirs at home

The topic was chosen, since the environmental perception analysis (▶ 9) showed that only 22.5% of the interviewees from Brumal had access to treated water. Users without access could be inadvertently exposed to water-borne diseases and As contamination. The workshop dealt with the importance of water for human consumption (origin, cycle and uses), with water purification needs and options (problems of home water storage, including the role of accumulating sediments in the storage containers), and with concrete guidelines on how to treat drinking water and wash food. Audio-visual aids helped to demonstrate the correct procedure for cleaning water reservoirs, the importance of filtering drinking water and of removing sediments from the bottom of the reservoirs, especially during the wet season (Figure 15.6).

The participants learned about the As remains sorbed to iron and manganese hydroxides in the sediments. In the rainy season, the amount of sediment carried to the water reservoirs increases the As content (Fig. 15.6). Given that the water distribution system does not work 24 hours a day, every time the system is filled, sediment is stirred up and stays suspended for hours. To drink this water may substantially increase involuntary As ingestion. The participants were interested and decided to clean their water reservoirs more frequently. Informative booklets on cleaning private water reservoirs were distributed by COPASA.

5. Education on water usage

Since the community of Santana do Morro would start using treated water supplied by a new community water treatment plant (▶ 16), this topic was selected to raise awareness on the importance of potable water quality and water use. A multimedia show presented information on water conservation and daily actions of an individual (Fig. 15.7). The function and importance of the new plant, maintenance requirements, and the necessity of protecting water as a resource were given priority. The Minas Gerais Sanitation Company COPASA was included and concrete advice given on how to store treated water, how to clean domestic water, and how to avoid water waste and leakage. The workshop closed with reading aloud a "prayer for water" (*Oração da água*) by everyone present – a moment of powerful emotion. The participants received educational material about water and treated water by COPASA. Treated water samples were distributed, since most participants had never had good quality water before. This was a moment of happiness, euphoria and curiosity for everyone.

6. Preventing As contamination through environmental dust

The theme was defined by results from the environmental perception study (▶ 9), where 32.2% of the interviewees in Barra Feliz reported living without paved roads and that 57.6%

Figure 15.6.   Left and center: water collection from home reservoirs. Right: loaded filter with residues and fresh filter (photos Michelly Amorim).

Figure 15.7.   Lecture for the community of Santana do Morro at the Residents Association (photos Michelly Amorim).

of their children played in the dust (dirt or unpaved ground). In Brumal, 19.7% were without pavement, and 35.0% of the children played on potentially As contaminated surfaces. Since As exposure through contaminated soil in these townships proved significant (▶ 12), the workshop was aimed at raising public awareness of local As contamination and exposure pathways, explaining preventative action (▶ 17), and presenting results from the urine study (▶ 14). Using common general knowledge about As toxicity and exposure pathways, people were given hands-on ideas on how to minimize contact with potentially contaminated material. A significant component was dedicated to reducing hand-mouth contact of little children and simple personal hygiene, e.g., washing hands before eating or handling food items. It was also demonstrated how house cleaning can be used to both reduce As-related risks (dust) and kill germs at the same time. The most demanding part was communicating the urine test results. Since the abstract numbers were not very meaningful to barely educated villagers, these parents were approached in an honest, sincere, clear and objective manner, and reassured, since none of their children presented As levels above safe limits.

The approach for this theme in the workshop for the pre-school children matched their age and level of understanding. The activity was developed in a very playful and relaxed way. The other participants also expressed great interest and willingness to follow the advice given to them (Figs. 15.8 and 15.9).

7. Garbage – its impact on environment, health, and a source of income
The topic was defined by results from the environmental perception study (▶ 9), where 92.4% of the dwellers reported having access to garbage collection services and 23% reported the existence of "alternative practices", such as garbage incineration. Analyzing the data per district and township, it was found that the people of the Santa Bárbara district (Barra Feliz, Brumal and Santa Bárbara town), besides presenting the lowest coverage for garbage collection services, make the most frequent use of practices like garbage incineration (e.g., 45.1% in Brumal). Related awareness was to be built, and information was disseminated about the importance of integrated urban solid waste management and garbage recycling to prevent and correct negative impacts on environment and health. The one-hour lecture "Can garbage be a source of income?" was born from the idea to update the public on the concepts of solid waste (definition, types, origin) and the impacts of this waste on the environment and health. People were also informed about the regional "Garbage and Citizenship Forum" (*Fórum Lixo e Cidadania*) and the importance of managing wastes properly. People were also informed about the multiple advantages of selective garbage collection. Apart from saving resources by increasing recycling rates, this system helps by including people who scavenge litter bins for recyclables, thus adding to a sustainable approach. Everyone showed interest in the topic, demonstrated by enthusiastic discussions and active participation in playful garbage separation quizzes and exercises (Fig. 15.10).

Figure 15.8.   Workshop for parents of students of Laudelina Antônia Gonçalves Municipal School, Barra Feliz (photos Michelly Amorim).

Figure 15.9.   Workshop for pre-school children of Cecília Álvares Duarte Municipal School (photos Michelly Amorim).

Figure 15.10.   Urban solid waste collected for the workshop (photos Michelly Amorim).

8. Waste incineration and the greenhouse effect

Results from the environmental perception study identified waste incineration as one of the environmental problems affecting communities and hence, defined this topic (▶ 9). Following up on the previous workshop topic, impacts of open air garbage burning on the environment and human health, particularly in relation to the greenhouse effect, were discussed. The greenhouse effect was explained to help people understand the larger concept of climate change. The workshop started with updating the participants on topics related to garbage, such as: concept, types, origin (sources), chemical composition, diseases caused by the poor handling of garbage; and the reasons why we should be committed to the correct disposal of solid waste. The most relevant aspects of the topic were addressed, such as the impacts

Figure 15.11.   Workshop for students and educators in Brumal (photos Eleonora Deschamps).

of garbage burning on the environment and human health, and the prospects of finding a solution to a socio-environmental problem of the community of Brumal.

The principles of the *Letter from Tbilisi*[1] served as a motivation to the educators and encouraged them to value individual action and its importance to the collective, thus correlating the local aspects (open air garbage incineration) to the global issue (greenhouse effect). The approach was similar for 1st-year and 4th-year elementary school students, after making adjustments for their understanding and age. The atmosphere of the workshops was very relaxed (Figure 15.11). All participants received copies of an informative booklet describing how to sort urban solid waste (FEAM 1995). Teachers received an additional booklet on practical approaches to air quality preservation, to be used in their didactic activities.

15.4   WORKSHOP EVALUATION

The experience in environmental education was challenging and gratifying at the same time. Working in the communities presented the team with unique and surprising moments, that were sometimes deeply emotional. The ARSENEX project was accepted by the communities from the very beginning. This was a major support for the work in the education workshops, whether these were held in schools or in community centers. Various community members became friends with team members – from children to the elderly. Team members were often recognized on the road and addressed by name and with affection. Often, people would approach and ask about the progress of the project, and when the team would be revisiting to their village. A most memorable example occurred in Santana do Morro with the construction of the water treatment plant (▶ 16). From the very beginning, the residents were eager to know where and how the station would be built, when it would be ready and when they would start receiving treated water. The project's initial aim earning the confidence of the people and having some familiarity with them, was achieved, not only during sample collection but also during the workshops.

During the workshops, the interest, presence, curiosity and enthusiasm of the participants was strongly felt. It was truly rewarding to contribute by spreading knowledge on arsenic and health-related issues. The opportunity was used to expand the scope to other environmental issues, such as garbage incineration and the use of untreated water. Consequently, changes in attitude and behaviour of the population that will attenuate local environmental problems can almost be taken for granted. The authors believe that the knowledge gained

---

[1] Declaration of the Intergovernmental Environmental Education Conference, Tbilisi, 1977.

by the participants and the public awareness campaign during the project will enable an ever-larger number of people to act and minimize the socio-environmental problems in the project area.

REFERENCES

Brasil: Law n° 9.795, from April 27, 1999 on environmental education on the federal and provincial levels. Diário Oficial da União – D.O.U.; Poder Executivo, de 28 de abril de 1999. Brasilia, 1999.

Carson, R.: *Silent spring*. Reprint 2002, Houghton Mifflin Books, New York, 1962, p.404.

Council on Environmental Quality: *The global 2000 report to the President of the US entering the 21st Century*. Pergamon Press, New York, 1, 1980, p.360.

Dias, G.F.: *Educação ambiental: princípios e práticas*. 9th ed.; Gaia editora, São Paulo, 2004, p.551.

FEAM: Como destinar os resíduos sólidos urbanos/Fundação Estadual do Meio Ambiente, Belo Horizonte. Série manual 1, 1995, p.45

Meadows, D., Meadows, D.L., Rander, J. & Behrens III, W.W.: *The limits to growth*. Universe Books, 1972, p.205.

Obara, A.T., Kiouranis, N.M.M. & Silveira, M.P.: Oficina de educação ambiental: desafios da prática problematizadora. In: Enseñanza de lãs Ciências. *VII Congreso Internacional sobre investigación en la didáctica de las ciencias*. 2004, p.15.

Oberdá, S., Deschamps, E. & Fittipaldi, L.: Educação ambiental e em saúde. In: Deschamps, E., Matschullat, J. (eds): *Arsênio antropogênico e natural. Um estudo em regiões do Quadrilátero Ferrífero*. Fundação Estadual do Meio Ambiente (FEAM), Belo Horizonte, 2007, pp.273–299.

Oliveira, V.: Educação Ambiental: uma proposta curricular. Tecnologia Educacional, Rio de Janeiro, 17/18, n. 85/86, 1988, 1989, pp.37–47.

Porto, M.F.M.M.: Educação ambiental: conceitos básicos e instrumentos de ação. Belo Horizonte: Fundação Estadual do Meio Ambiente; DESA/UFMG, 160; (Manual de Saneamento e Proteção Ambiental para os municípios), 1996.

ADDITIONAL MATERIAL, NOT CITED IN THE TEXT

Cadernos de Inovação Pedagógica: O meio ambiente e a escola. Editora Viva n° 2. DIVIP – MG http://portal.mec.gov.br/secad/arquivos/pdf/educacaoambiental/historia.pdf, 1995.

Cadernos Periódico: Um pouco da história da educação ambiental. http://portal.mec.gov.br/secad/arquivos/pdf/educacaoambiental/historia.pdf.

Chemello, T.: *Brincando com Embalagens Vazias*. Global Editora, 2010, p.68.

Coletto, L.M.M.: Manual Informativo – Programa de Educação Ambiental – Superintendência dos Recursos Hídricos e Meio Ambiente – SUREHMA - A Qualidade da água e sua importância para a Vida. 3rd ed., Curitiba, 1989, p.19.

Coordenação de Educação Ambiental, M.E.C.: A implantação da educação ambiental no Brasil; Conceitos para se fazer educação ambiental. http://www.dominiopublico.gov.br/pesquisa/DetalheObraForm.do?select_action=&co_obra=24736, 1998.

COPASA: Saneamento – uma questão de saúde, desenvolvimento social e econômico. Companhia de Saneamento de Minas Gerais, Belo Horizonte, no year.

Czapski, S.A.: Implantação da educação ambiental no Brasil. Brasília: Ministério de Educação e do Desporto, 1998, p.166.

Freire, P.: *Educação e Mudança*. Rio de Janeiro: Paz e Terra, Rio de Janeiro, 1979, p.46.

Freire, P.: *Pedagogia do oprimido*. 19th ed. Paz e Terra Editora, Rio de Janeiro, 1991, p.107.

Guimarães, M.A.: *A dimensão ambiental na educação*. 8th ed: Papirus Editora, Campinas, SP, 1995, p.107.

Jovens em Ação: *Ações para melhorar o ambiente e a qualidade de vida nas cidades*. Editora Melhoramentos SP, 2000.

Marcatto, C.: Educação Ambiental: conceitos e princípios. Belo Horizonte, FEAM, 64, 2002.

Marcos, D.: *Natureza da Paisagem. O lixo pode ser um tesouro*. Cima, 1999.

Medina, N.M.: Os desafios da formação de formadores para a educação ambiental. In: Junior, A. & Pelicioni, M.C., (eds) *Educação Ambiental: desenvolvimento de cursos e projetos.* USP, São Paulo, 2000, pp.9–27.

Prefeitura Municipal de Belo Horizonte: Pé de moleque – Cartilha Educativa, 1997.

Programa Ambiental: A Última Arca de Noé. Educação Ambiental e o poder público. http://www.aultimaarcadenoe.com/podereduca.htm.

Schall, V.: *Ciranda do meio ambiente, concepção e coordenação.* Editora Memórias Futuras, Rio de Janeiro, 1991.

Seara Filho, G: Apontamentos de introdução à educação ambiental. *Revista Ambiental* 1: 1 (1987), pp.40–44.

Vieira, L.: Metodologia de educação ambiental para indústria. Contagem, Santa Clara, 2004, p.168. http://www.ldes.unige.ch/bioEd/2004/pdf/vieira.pdf.

CHAPTER 16

Water treatment – A local example

Eleonora Deschamps & Neila Assunção

16.1 INTRODUCTION

There has always been a close connection between the location of human settlements and drinking water sources. More recently, growing awareness of the importance of water quality emerges. In spite of this, many children die each year from diseases related to insufficient or contaminated water supply as well as sewage problems. The World Health Organization (WHO) estimated that one child died every eight seconds from water-related infectious disease in 1996. Heller and Pádua (2006) pointed out that Brazil must provide all citizens with safe drinking water in order to protect their health and to ensure a sustainable relationship with the environment.

No single solution exists to supply people with water. Economic, social and operational factors need to be considered when making related decisions. The best water supply system therefore is neither necessarily the most modern, nor the most economic, but one that fits the socio-economic reality of the locale. Often, two or more alternatives have to be compared, based on cost-benefit analysis. UNICEF (1978) defined appropriate sanitation technology as being:

- **Hygienically safe**: prevents spreading diseases and encourages healthy habits, risk-free
- **Technically and scientifically satisfactory**: user-friendly with easy maintenance, technically efficient and effective, sufficiently adaptable to variable situations
- **Socially and culturally acceptable**: meets the basic needs of the population, makes use of local labor, improves and does not replace – as far as possible – traditions, attitudes and crafts, aesthetically pleasant
- **Innocuous to the environment**: curbs environmental contamination, maintains ecological balance, contributes to the conservation of non-renewable resources, re-circulates sub products and residues, enriches rather deprecates the environment
- **Economically feasible**: is cost-efficient, preferably adopts inexpensive solutions. Contributes to the development of local industry, uses local materials and uses energy economically.

Among the key parameters to be addressed when planning a water treatment plant are: the size of the target community, resources needed, quantity and quality of water to be consumed, topographic and climatological characteristics, installation options with help from local community members, public funding, availability of electric power and of human specialist and financial resources. The largest demand for good quality potable water exists in localities with a small number of inhabitants, generally in the rural areas (Heller and Pádua 2006).

Santana do Morro (Santa Bárbara district, Brumal township) has about 200 inhabitants, 24 of whom are illiterate (12%). The locality lacks basic sanitation and is deprived of access to health and hygiene services. Water comes from a local spring without any treatment, and streets are unpaved. Water samples from this spring yielded As concentrations from 3.3 to 21.5 µg L^{-1}, regularly exceeding official potability standards (Table 16.1, ▶ 5). All water samples contained substantial levels of coliform bacteria, making the water generally unsuitable for human consumption (▶ 11, Tables 16.1 and 16.2). Sediments in those waters presented a mean value of 130 mg As kg^{-1}, again exceeding the standards (▶ 5, 12). The high

As concentrations in sediments additionally justified providing a water treatment facility (▶ 12, 15). Prevalent As species should be identified in order to make a decision on oxidation steps prior to the treatment. Here, no speciation-analysis was done, based on the experience that most, if not all, arsenic occurred in the fully oxidized form. The values refer to total arsenic, since the catchment favors oxidation and the manganese oxide containing sorbent further promotes As oxidation to $As^{(V)}$.

Only colour and fecal coliform bacteria exceeded the limits established by the Ministry of Health in the Brumal region (Table 16.1; Brasil 2004). Mean values do not reflect reality, however. Monthly analytical monitoring showed that the threshold values were exceeded ten-fold, particularly in the rainy season. Turbidity, coliform bacteria and iron levels regularly exceeded the limits in Santana do Morro, confirming poor water quality. Table 16.2 delivers an estimate of the quantity of microorganisms that cause diseases (Di Bernardo 1993 in Heller 2006), albeit with substantial individual variations (receptor sensitivity).

This situation motivated planning for an economically feasible water purification plant within the scope of the ARSENEX project to serve 50 households and improve their quality of life. COPASA initiated field inspections together with some project team members to identify the best location for the future plant. Following assessments of topography and water availability and the existing structures in the locale, the team decided on a location known as "Mr. Nilo's spring". An area of 190 m^2 was then formally granted to Santa Bárbara City Hall by a community resident to install the plant. The area around the spring was fenced to prevent animal access.

Sand and gravel filters were selected as a simple and safe cost-cutting method to treat water by combining physical, chemical and biological processes to comply with the quality parameters (e.g., colour, turbidity, dissolved solids and coliform bacteria) as demanded by law. Although it is an old technique, invented in the 19th Century in the United Kingdom (Di Bernardo and Di Bernardo-Dantas 2005), it has proven its reliability and applicability particularly in small rural properties (Di Bernardo et al. 1999 in Murtha and Heller 2003). Sand filters were used for the first time to supply London with water in 1828. Between 1914

Table 16.1.   Potable spring water quality in Brumal and Santana do Morro (values rounded).

Parameter	Region of Brumal variation[a] (mean[a])	Santana do Morro	Administrative rule 518/04
pH	5.9–7.9 (7.0)	7.7	6 to 9.5
Turbidity (NTU)[1]	0.08–14 (2.0)	6.5	2
Fecal coliforms (CFU 100 mL^{-2})	0–2400 (148.7)	3	Absent
Colour (CU[3])	1–180 (30)	1	15
Soluble iron (mg L^{-1})	0.01–0.31 (0.16)	1.0	0.3
Total manganese (mg L^{-1})	0.01–0.25 (0.13)	0.01	0.1

[1]NTU: nephelometric turb. unit; [2]CFU: colony formation unit; [3]CU: colour unit; [a]CVRD (2001).

Table 16.2.   Quantity of bacteria that may trigger a disease (estimate by Di Bernardo 1993).

Organism	Quantity	Disease
Shigella dysenteriae	10	Dysentery
Vibrio colerae	1000	Cholera
Salmonella typhi	10000	Typhoid fever
cysts of Entamoeba histolytica	20	Amoebian dysentery
Escherichia coli	1×10^{10}	Gastroenteritis

and 1918, this method, known as slow filtration, was replaced with the quick filtration process, since the latter can treat larger volumes of water per area (Bolmann 1987 in Londe 2002). From the 1950s onwards, slow filtration again started to attract attention, based on new studies seeking a simple and effective water treatment technology for the United States and some European countries (Valenzuela 1991). In 1980, slow filtration was responsible for the treatment of 28% of the water in the United Kingdom. In individual parts of England, more than 70% of the water was treated by the slow method (Costa 1980; Hespanhol 1969; Mbwette and Graham 1987 in Londe 2002). Today, this technology is again used in cities like London (80% of the treated water supply). In Brazil, conventional treatment and direct filtration predominated, while techniques like slow filtration, flotation and membrane filtration were used in relatively small numbers. Table 16.3 compares the slow and the conventional filtration process.

Slow filtration is cheaper to implement and the operational costs are much lower than the conventional treatment method. This makes slow filtration more acceptable to small communities. The structure of a water treatment station can be built of ferrocement (Fig. 16.1), the most appropriate building material for water reservoirs and tanks (Ferrocement 2010). It consists of cement-rich mortar, reinforced with a wire or steel mesh. This low-cost technique (about half the price of other technologies), is frequently used in rural environments.

A spillway upstream from the slow filter homogenizes the flux to the plant inlet, and improves the efficiency of one or more pre-filters that provide an initial influx treatment. The pre-filtration step increases the efficiency of the slow filtration (Di Bernardo and Di Bernardo-Dantas 2005). Pre-treatment is needed when the concentration of suspended solids in raw water is high (Heller and Pádua 2006). Turbidity of the water effluent to the slow filter may not exceed 10 NTU (Table 16.4).

Table 16.3. Characteristics of slow filtration and conventional treatment.

Parameter	Slow filtration	Conventional treatment
Operation	Simple	Complex
Coagulant consumption	None	High
Resistance to variations in water quality	Low	High
Filter cleaning	Scraping of surface layer	Ascending flow
Station size	Limited to small	No limits
Implementation costs (US$ per person)	10–100	10–60
Area needed	Large	Medium

Figure 16.1. Construction with ferrocement (photos Neila Assunção).

Table 16.4.   Recommended water parameters for slow filtration.

| Parameter | Recommended value by authors | | | |
	Spencer and Collins (1991)	Cleasby et al. (1984)	Di Bernardo (1993)	Heller (1992)
Turbidity (NTU)	10	5	10	10
Algae	20×10^5 ind. L^{-1}	5 µg chlorophyll a L^{-1}	2.5×10^5 ind. L^{-1}	–
Colour (CU)	25	–	5	5
Total iron (mg L^{-1})	1	0.3	2	1
Manganese (mg L^{-1})	–	0.05	0.2	0.2
Fecal colif. (NMP 100 mL^{-1})	–	–	200	500

Galvis et al. (1997) in COPASA (2004).

The slow filter is formed by a sand layer (0.90–1.20 m) on top of one or more gravel layers (0.20–0.45 m). It has an inlet for raw water and an outlet for treated water. The water continuously changes direction while flowing through the filter, thus optimizing the contact of impurities with the filter grains (Di Bernardo and Di Bernardo-Dantas 2005). One of the main advantages of the slow filter is its efficiency in removing bacteria and viruses, based on the natural sloping ground, internal predation, the biocidal effect of solar radiation, and the sorption of these organisms to the biofilm.

Suspended material affects the operation up-time, reducing the bed volumes prior to filter cleaning (▶ 3, 4). The time span between filter maintenance tends to exceed two months and may be much longer, if the raw water carries little dissolved and suspended matter. Sand characteristics affect the biofilm formation time, too (Di Bernardo and Di-Bernardo-Dantas 2005). This biofilm above the sand layer [Schmutzdecke in German, the "dirt layer"] is formed by inert particles, inorganic matter and bacteria, algae, protozoa, invertebrates and their extra-cellular products, besides iron and manganese, when these materials meet in the effluent (Di Bernardo and Di-Bernardo-Dantas 2005). The removal of Fe and Mn in slow filtration occurs by precipitation of these metals in the biofilm; an excessive load may cause quick colmation of the sand bed. The presence of algae in the effluent may have a negative impact on the treatment, since high algal concentrations may clog the flux, demanding more frequent cleaning.

The water level in the slow filter ranges from a minimum at the beginning of the filtration, to a maximum pre-established value, when filter operation must be interrupted for cleaning. The latter consists of mechanically scratching a surface layer (1–3 cm). This sand layer must be restored when it reaches a thickness of 60 cm after successive cleaning operations. The same sand is scratched from the surface and can be re-used after proper cleaning. Disinfection is a process in which a chemical or non-chemical agent is used to inactivate pathogenic microorganisms in the water (Di Bernardo and Di Bernardo-Dantas 2005).

16.2   TECHNICAL DETAILS OF THE TREATMENT PLANT

Raw water for the Santana do Morro plant is taken from the dammed (20 m) and fenced source of Santo Antonio creek. Water is led via 100 m of PVC DN 100 tubes to the plant, and another 183 m of PVC DN 50 tubes and 474 m of PVC DN 32 tubes were built into the system. The flow rate was calculated, taking local demand and the treatment station capacity into account. The plant life time was estimated with 20 years, the coefficient for daily peak consumption (k1) as 1.2, that for peak consumption per hour (k2) as 1.5 and a per capita consumption of 150 L day^{-1}. The resulting flow is 0.76 L s^{-1} (65.66 m^3 day^{-1}).

The roughing filter (2 m diameter, height 2.45 m) was built of ferrocement (Fig. 16.1). Three gravel layers (30 cm each) of decreasing granulometry (19–31, 6.4–12.7, and 2.4–4.8 mm); and a bottom layer of 36 cm (1.4–2.0 mm) form the filter for vertical water ascent. The slow filter (ferrocement) was built downstream of the roughing filter (4.00 m diameter, 3.05 m high). The supply rate is 6.0 m^3 m^{-2} day^{-1}, and the calculated filter maturation time is 15 days. The filter bed is formed by a 60 cm layer of broken stone and four gravel layers (17 cm: 17–38 mm, 10 cm: 7.9–15.9 mm, 8 cm: 3.2–6.40 mm, and 5 cm: 1.4–2.6 mm). The slow filter contains a 90 cm sand layer, on top of which is a water column (85 cm). The disinfection unit was made of ferrocement, too. Sodium hypochlorite is continually added into the contact tank with a storage capacity of 5 m^3 and a minimum retention time of 30 minutes. Two pumps with 2 kV each transport the water from the contact tank into the As filtration units – one with synthetic filter material, the other one with a natural sorbent.

The synthetic filter with GEH (Granular Ferric Hydroxide®, donated by the German producer) consists of granulated synthetic iron hydroxide from high-purity materials (▶ 3, 4). The mean porosity of the filter is 75% and the specific surface area measures 250–300 m^2 g^{-1}. GEH® is a high-performance sorbent that does not leave dangerous residues in the water, nor affects the pH. Its capacity to sorb arsenic and phosphate reaches 60 g kg^{-1}. After saturation (about 36 months), the filter material should be disposed of in a landfill. This is not a feasible solution for the community, since such periodic replacement adds unbearable costs for the people. A solution to this problem was sought with natural low-cost sorbents, originating from capstones of an old iron mine in Santa Bárbara (Deschamps *et al.* 2003, 2005). This natural sorbent, known as cFeMn-c, basically comprises of iron (goethite, hematite and magnetite) and manganese minerals (todorokite, birnessite, cryptomelane and lithiophorite; Table 16.5). The sorbent (particles between 1 and 4 mm) is compressed in a bed and works like a conventional granulated activated charcoal filter. Arsenic is removed through oxidation (As$^{(III)}$ to As$^{(V)}$) by manganese and subsequent sorption and co-precipitation with Fe-phases. This material was applied for the first time on a practical scale and with excellent test results. The GEH® filter was installed to act as a backup for the local material, since GEH® efficiency and reliability have been demonstrated for long.

After As removal, the water flows to the storage unit (ferrocement) with a capacity of 30 m^3. The water is distributed by gravity with 1.15 L s^{-1}. A total of 1,516 m of tubing were installed (140 m PVC DN 40, 50 mm; 225 m DN 32, 40 mm; 1,151 m DN 25, 32 mm). Inside the filters, the filling heights for the filtration materials were marked to ensure the correct thickness of each layer as designed in the plan, and to avoid the trouble of having to remove the material to correct the thickness of the layers (Fig. 16.2). While such steps may appear irrelevant to professionals, they are of key relevance when working with local laypeople.

Table 16.5. Mineralogical composition of cFeMn-c.

Mineral	Weight-%
Goethite (FeO[OH])	30
Hematite (αFe$_2$O$_3$)	19
Todorokite (Ca,Na,K)(Mg,Mn$_2$+)Mn$_5$O$_{12}$.xH$_2$O)	6
Lithiophorite (AlLiMn$_2$(OH)$_2$)	2
Cryptomelane (K$_2$-xMn$_8$O$_{16}$)	15
Birnessite (Na,Ca,K)(Mg,Mn)Mn$_6$O$_{14}$.5H$_2$O	13
Quartz (SiO$_2$)	7
Muscovite (K,Na)(Al,Mg,Fe)$_2$(Si$_{3.1}$ Al$_{0.9}$)O$_{10}$(OH)$_2$	8

Deschamps (2003).

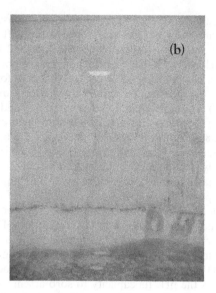

Figure 16.2.   Marking filter material levels inside the filter tank walls: (a) roughing filter, (b) slow filter (photos Neila Assunção).

Another very important task was sieving the filter materials to obtain the specified granulometries. The materials must be free of organic and inorganic impurities, and re-washed, if necessary. After marking the filter walls and separating the material to the specified granulometry, the filter materials were placed inside the units. Thereafter, influx was regulated and the filters filled. If this flux is below the demand, water shortages would result, which may lead people to seek other water sources – most likely untreated. Should the effluent flux be above demand, water quality will be compromised, since excessive transit speeds through the filter layers render the physico-chemical and biological processes infeasible. While the population would receive plenty of water, its quality is compromised due to improper treatment.

16.3   START-UP AND PLANT PERFORMANCE CONTROL

The project team monitored the works and the final start-up of the water treatment plant in Santana do Morro. Regular on-site visits to collect water samples and measure basic parameters (turbidity, pH, coliform, and colour) of raw and filtered water, and analyses COPASA in Santa Bárbara helped controlling the operation. The complete plant is shown in Figure 16.3.

The effluent flow was adjusted until it reached the calculated demand (0.76 L s^{-1}; 65.66 m^3 day^{-1}). This control was done by measuring the water volume in the spillway within a pre-determined period of time (Table 16.6). It is very important to check the performance of the spillway, since any deviation from the design would compromise the plant's efficiency. The performance of the triangular spillway is checked by measuring the height (h) of the water column from the bottom of the triangle, and the reduction in water volume in a pre-determined period of time (Table 16.7).

The results confirmed that the plant met all design parameters. Following this verification and related adjustment, the filter materials were filled in. The biofilm on top of the filter bed forms naturally within days or weeks. In one of our first visits to the plant, the biofilm was already visible (Fig. 16.4). The greenish colour results from the algae on the water.

Figure 16.3.   Plant units in Santana do Morro; the As filters at upper left (photos Neila Assunção).

Table 16.6.   Affluent flow.

Volume (L)	overflow	5.5	4	3.5	3
Time (s)	10	5	5	5	5
Flow (L s⁻¹)	not determined	1.1	0.8	0.7	0.6

Table 16.7.   Water level × volume × time × flow into the spillway (values rounded).

Height (cm)	Volume (L)	Time (s)	Flow (m³ s⁻¹)
2.0	3.5	5	0.20
2.4	6.0	3	0.30
3.0	0.7	3	0.49
3.5	1.4	3	0.68
5.0	1.3	3	1.48

Filters need to be cleaned on a regular basis by scraping a layer of sand off the surface to prevent clogging (colmation). Figure 16.5 shows a recently cleaned filter. An unusually high water flow that could reach the slow filter during plant operation would force a depression into the filter medium. This would trigger the formation of canals that prevent proper filtration (by-pass effect). To minimize such effects, the incoming water can be distributed using perforated pipes. These distribute the water equally over the filter surface, with lower pressure, like a shower (Figs. 16.4 and 16.5).

16.3.1   Analytical results

Monitoring a water treatment plant entails the verification of those parameters to ensure they comply with the regulating law (Table 16.8).

The data demonstrate that the water was initially unsuitable for human use. The increases turbidity and pH-values in the filtered water had been expected, since the unit was at the beginning of its operation. Then, filter media are not yet stabilized and the biofilm not yet established (Heller and Dantas 2006). Visits to the plant in Santana do Morro continued until all parameters were in compliance with Administrative Rule 518/04. Officially inaugurated in June 2007, the water treatment plant has been monitored on a regular basis by Santa Bárbara City, operates well and is of great benefit to the people of Santana do Morro (▶ 17).

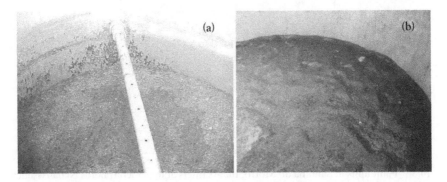

Figure 16.4.   Biofilm on filter media: (a) coarse pre-filter (b) Slow filter (photos Neila Assunção).

Figure 16.5.   Filter after cleaning (photos Neila Assunção).

Table 16.8.   Analytical results in the initial operating phase.

Parameter	Raw water	Filtered water	Adm. rule 518/04*
Colour (CU)	40	90	15
Turbidity (NTU)	0.79–0.87	1.3	2
pH	6.6	7.3	6–9.5
Total coliforms	present	present	none
Fecal Coliforms	present	present	none

*Brasil (2004).

REFERENCES

Bollmann, H.A.: *Aplicação da filtração lenta na remoção de substâncias contidas em águas superficiais.* Unpubl. M.Sc. thesis USP-EESC São Carlos, 1987, p.207.

Brasil: Ministério da Saúde. Portaria n. 518, de 25 de março de 2004. Ministério da Saúde, Secretaria de Vigilância em Saúde, Coordenação Geral de Vigilância em Saúde Ambiental, Brasília, Editora do Ministério da Saúde, 2005, 2004, p.28.

Cleasby, J.L., Hilmoe, D.J. & Dimitracopoulos, C.J.: Slow-sand and direct in-line filtration of a surface water. *J AWWA* 76:12 (1984), pp.44–55.

COPASA: Projeto básico do sistema de abastecimento de água. Santa Bárbara – localidade de Santana do Morro/MG. Superintendência de gerenciamento de obras. Divisão de gerenciamento de obras do interior; Belo Horizonte; Single volume, 2004, p.68.

Costa, R.H.R. da: *Estudos comparativos da eficiência de filtros lentos de areia convencional e de fluxo ascendente.* Unpubl. M.Sc. thesis USP-EESC São Carlos; 1980, p.256.

CVRD: Boletins de monitoramento da qualidade das águas – área de influência de Brumal. March to May, September, November and December 2000 and March to May 2001. Companhia Vale do Rio Doce, 2001.

Deschamps, E.M.: *Avaliação da contaminação humana e ambiental por arsênio e sua imobilização em óxidos de ferro e de manganês.* PhD thesis, Escola da Engenharia, Federal University of Minas Gerais, Belo Horizonte: 2003, p.139 + annex.

Deschamps, E., Ciminelli, V.S.T., Weidler, P.G. & Ramos, A.Y.: Arsenic sorption onto soils enriched with manganese and iron minerals. *Clays Clay Minerals* 5:2 (2003), pp.197–204.

Deschamps, E., Ciminelli, V.S.T. & Höll, W.H.: Removal of As[III] and As[V] from water using a natural Fe and Mn enriched sample. *Water Res.* 39:20 (2005), pp.5212–5220.

Di Bernardo, L.: *Métodos e técnicas de tratamento de água.* Vols. 1 and 2. ABES, Rio de Janeiro, 1993.

Di Bernardo, L., Brandão, C.C.S. & Heller, L.: Tratamento de águas de abastecimento por filtração em múltiplas etapas. PROSAB – Programa de Pesquisa em Saneamento Básico. ABES Associação Brasileira de Engenharia Sanitária e Ambiental, Rio de Janeiro, 1999, p.114.

Di Bernardo, L. & Di Bernardo Dantas, A.: *Métodos e técnicas de tratamento de água.* 2nd ed; RiMa Editora São Carlos, 2005, p.792.

Ferrocement: web page www.ferrocement.com, 2010.

Galvis, G., Latorre, J., Fernandez, J. & Visscher, J.T.: *Multistage filtration. A water treatment technology.* IRC Cinara, Universidad del Valle. The Hague, Netherlands, 1997.

Heller, L.: Tecnologias de baixo custo para países em desenvolvimento. In: *5 Simposio Brasileiro de Engenharia Sanitaria e Ambiental.* Anais Lisboa, 1992, p.100.

Heller, L. & Padua, V.L. de (eds): *Abastecimento de água para consumo humano.* Editora UFMG, Belo Horizonte, 2006, p.859.

Hespanhol, I.: *Investigação sobre o comportamento e aplicabilidade de filtros lentos do Brasil.* Unpubl. M.Sc. thesis Universidade de Sao Paulo, 1969, p.133.

Londe, L., de, R.: *Eficiência da filtração lenta no tratamento de efluentes de leitos cultivados.* PhD thesis, Universidade Estadual de Campinas UNICAMP, Faculdade de Engenharia Agrícola, 2002, p.80 + annex.

Mbwette, T.S.A. & Graham, N.J.D.: Improving the efficiency of slow sand filtration with non-woven synthetic fabrics. *Filtration Separation* 24:1 (1987), pp.46–50.

Murtha, N.A. & Heller, L.: Avaliação da influência de parâmetros de projeto e das características da água bruta no comportamento de filtros lentos de areia. *Engenharia sanitária e ambiental* 8:4 (2003), pp.257–267.

Spencer, C.M. & Collins, M.R.: Water quality limitations to the use of slow sand filters. *Slow sand filtration Workshop. Timeless technology for modern applications,* 1991.

UNICEF: Primary health care: *Report of the international conference on primary health care.* Alma-Ata, USSR; Geneva, 1978.

Valenzuela, M.G.R.: *Estudo do desempenho de uma instalação de pré-filtração lenta com mantas para o tratamento de águas de abastecimento.* Unpubl. M.Sc. thesis Escola de Engenharia de São Carlos – USP, São Carlos; 1991, p.417.

Vareche, M.B.: *Estudo sobre a interferência de algas no sistema de filtração lenta em areia.* Unpubl. M.Sc. thesis Escola de Engenharia de São Carlos – USP, São Carlos, 1989.

CHAPTER 17

Mitigation measures and solutions

Eleonora Deschamps & Jörg Matschullat

17.1   RISK EVALUATION

To assess risks related to the As anomalies in the Iron Quadrangle (ARSENEX project) and
to deliver a quantitative explanation for elevated As values in human urine (▶ 14), all potential As pathways were investigated, based on the experience gained in evaluating the results
from environmental compartments (▶ 10–14).

Data on ingestion of collard-green and lettuce showed that As uptake by children ranged
between 1 and 12% of the provisional tolerable weekly intake (PTWI) in the Santa Bárbara
district, and for adults between <1 and 4% of the PTWI. Values were considerably higher
in the Nova Lima district: adults ingested between 16 and 72%, and children from 55 to
250% of the PTWI. Even with a conservative estimate, the encountered body As load can
be explained by this single exposure pathway. Food was not the only contamination source.
While As-contaminated water appeared not to be a serious issue (▶ 11) as long as people did
use recommended sources, soil and dust are likely As exposure sources at least in some places
(▶ 12). Children are potentially more exposed to environmental contaminants due to their
intrinsic social and behavioral nature. Children younger than 7 years tend to be more in contact with soil, water and biota, keeping frequent manual contact, ingesting and inhaling it.
The data on the soil show that dust is an exposure pathway of the same order of magnitude
as As ingestion via food (▶ 10; Table 17.1).

Table 17.1.   Total As exposure in the Nova Lima and Santa Bárbara districts.

	Nova Lima	Santa Bárbara	"Normal"*
	Oral absorption (μg As day^{-1} person^{-1})		
Food	24–110	0.5–16	10
Water	0.2–4	<0.1–2	1.4–20
Soil			
Children	10–20	<<1	<<1
Adults	1–2	<<1	<<1
	Inhalation (μg As day^{-1} person^{-1})		
Children	1.5	0.4	0.03
Adults	6	1.5	0.1
	Total As uptake (μg As day^{-1} person^{-1})		
Children	40–136	3–18	10–30
Adults	35–122	4–20	
	Percentage of PTWI		
Children	92–316	7–43	26–70
Adults	23–81	3–13	7–20

*The "normal" values were compiled from various sources (Rassbach 2005).

The concentration of As compounds in the environment is not equal to the amount sorbed by the human organism (▶ 2, 14). The exposure calculation suggests the As quantity that enters in contact with human organs, whether by inhalation, ingestion or sorption through the skin, thus enabling to link the exposure levels with possible effects on humans. With a carcinogenic substance, however, there are no safe exposure limits, and the entire population has to be considered as exposed to the hazard. Cancer is always rare and may appear long after the exposure (▶ 2). This explains why cancer, caused by exposure to potentially toxic chemical substances, is not diagnosed very often (▶ 9). It is difficult to detect true cause-effect relationships, since an adverse health effect may be the result of non-carcinogenic toxic exposures, or of the carcinogenic impact of a chemical toxin. Compiling all potential As exposure pathways, Table 17.1 presents the ranges of individual exposure in the Nova Lima and Santa Bárbara districts.

These results, calculated on the basis of more conservative figures, show a clear image of the risks in both districts. Many factors such as ingestion through food, inhalation of dust, and to a certain extent, uptake of As-enriched water contribute to the individual body As load. Risks are higher in the contaminated areas of Nova Lima, which corroborates the As concentrations found in the urine samples from that district. These results justified the preventative measures taken and the recommendations presented below. Although the risk for children is high, no direct investigation into arsenicosis has been undertaken yet. No clear evidence for arsenicosis was found during the ARSENEX project (▶ 14). It is important to note, however, that some children showed immunological disorders. This became obvious with frequent complaints about a whole range of infections (▶ 9). The risks are real, particularly for the children, and pointed to a pressing need of future monitoring the risk reduction measures, such as concrete seals for house floors in the Galo neighborhood and immobilization of the old tailings deposits to significantly reduce exposure pathway strength.

17.2   INSTITUTIONAL MEASURES

Institutional measures were implemented in the area in line with environmental and health education workshops (▶ 15). These were based on the results from the investigations in environmental and biological compartments (▶ 10–14) and as a result of the interaction between project team, private companies and governmental institutions.

Rehabilitation of the old tailings deposits of Nova Lima. Morro Velho Mining, today AngloGold Ashanti Mining, is a key industry in the Nova Lima district (▶ 6). As part of a "Conduct Adjustment Agreement", signed between the district attorney and the company, and mediated by the environmental authority FEAM in 2003, AngloGold started to recover the five old tailings deposits (▶ 11), areas where refuse of the metallurgical plant in Nova Lima had been deposited. This work started with the Galo deposit and consisted of an

Figure 17.1.   Galo deposit before (left) and after (right) rehabilitation (photo Jörg Matschullat).

extensive enclosure of the contaminated material. The structure was sealed with compacted clay layers and geomembranes, and a permanent drainage system was constructed to control contaminated water (Fig. 17.1).

The Resende deposit was the second one to be recovered similar to the Galo deposit (Fig. 17.2), followed by the Matadouro deposit, finished in 2008. Until 2010, all deposits have been recovered and incorporated into the city infrastructure.

Covering yards and vegetable gardens on the Galo deposit. Back and frontyards, and private gardens in the Galo neighborhood sat directly on top of the tailings materials. Following guidelines by FEAM and based on results from the environmental studies (▶ 10–14), all yards and gardens and soil, where plants were cultivated, were covered with concrete by AngloGold. Residents, who expressed the wish to keep their vegetable gardens, received a system of concrete boxes (elevated planters) to prevent direct exposure to the contaminated ground (Fig. 17.3). The alternative would have been to evict people from their homes – neither acceptable nor feasible.

In parallel, FEAM requested an epidemiological study under the coordination of the Nova Lima City Hall to better understand the possible impacts of As exposure on the health of the exposed residents (▶ 14). From 2007 to 2010, four new cases of cancer appeared among the residents of Morro do Galo, including two deaths. While such phenomena cannot be directly linked to the As exposure, it was recommended to continually monitor the exposed community, and to include this task in the public policy of the district. A short description of recommendations for ongoing measures follows.

Figure 17.2.   Resende deposit before (l.) and after (r.) rehabilitation (photos Michelly Amorim).

Figure 17.3.   Vegetable gardens in Galo Novo before (left) and after (center and right) yards and gardens were cemented (photos Jörg Matschullat).

17.3   RECOMMENDATIONS

For decision-making, for management and risk communication, it is important that the risk evaluation process is clear and transparent, and well known by all stakeholders (the population, especially exposed individuals, scientists and government, local authorities and companies), to guarantee the preservation of health and the quality of life of the population. Considering that the remediation measures, whatever their nature, cannot guarantee a total risk elimination, it was recommended to the public powers of the two districts covered in the ARSENEX project to:

- Establish an environmental education program for the population to deliver the knowledge needed for autonomous action to protect and promote their own health (▶ 15)
- Monitor the environmental quality to support the actions of vigilance for the health of the exposed population (▶ 10–14)
- Organize an information system to support the population (▶ 17)
- Monitor biological exposure indicators to follow and foresee health hazards, including a long-term cancer-risk related assessment (▶ 13, 14)
- Establish partnerships with health care and environmental institutions, and schools
- Promote the continuous qualification of health care professionals and community agents, to help them assist the actions geared at health surveillance of the exposed people (▶ 15)
- Continue the mitigation work along the exposure pathways (▶ 17)
- Develop studies and research to better explain the absence of chronic diseases and give a scientific approach to the intervention and decision-making process (▶ 2, 9, 14).

These recommendations have been implemented by now (2010). Their success does not lie exclusively in their formal implementation but rather in the support of both knowledge and of environmental and risk awareness of all stakeholders. The lesson learned to successfully reduce the described risks has various aspects, namely:

a) A sound scientific data base and related expertise to describe the potential risks and to develop and justify solutions,
b) A very open and pro-active information policy and the inclusion of all stakeholders in all steps of the process to improve dissemination of information on As exposure,
c) An intelligent and persuasive application of existing rules and regulations – and to control that they are being followed,
d) And last but certainly not least, the willingness of local residents to trust and collaborate with the project team. Without their engagement and adaptability to risk-reducing action, the ARSENEX project would have produced data and scientific results, but no true success.

While this may not always seem easy, and certainly requires dedication and perseverance, most people involved understand sooner or later, how successful risk reduction contributes to their quality of life and to a fruitful and trusting collaboration between the stakeholders. This much improves wellbeing and progress from the individual level to administration and government.

REFERENCES

Deschamps, E. & Matschullat, J. (eds): *Arsênio antropogênico e natural. Um estudo em regiões do Quadrilátero Ferrífero*. Fundação Estadual do Meio Ambiente, Belo Horizonte, 2007, p.330.
Raßbach, K.: *Arsentransfer und Human-Biomonitoring in Minas Gerais, Brasilien*. M.Sc. thesis, Unpublished, TU Bergakademie Freiberg, 2005, p.93 + annex.

Section V
Annex

Subject index

This index includes all keywords that deem helpful to the reader. It also provides direct access to those keywords within the references. Following a keyword, explicatory notes may appear in brackets. These are given to avoid misunderstandings, since much of the multidisciplinary terminology can be understood rather differently in individual fields of science (bio. = biology, chem. = chemistry, econ. = economics, eng. = engineering, geo. = geology, hist. = history, med. = medicine, min. = mineralogy, pharm. = pharmacy, soc. = sociology). Country names are given with the international abbreviation (ISO 3166 code). Chemical element names are listed with their abbreviation of the Periodic System of the Elements (IUPAC). Often used technical terms are written in full with the formal abbreviation thereafter, e.g., "acceptable daily intake (ADI)". Cross-references within the index are given with the keyword, followed by "*see* ...". We hope that you benefit from this index as a useful tool to rapidly access information and to get the best out of this book.

199

wallpaper (eng.) 3
waste (incineration), *see* garbage, refuse,
 sewage 7ff, 11ff, 22, 39f, 47, 51,
 60f, 66, 86, 96ff, 104f, 116, 119f,
 124, 126, 151, 173, 176ff
water (geo.), *see* bottled water, cloud water,
 drinking water = potable water,
 freshwater, groundwater, pore
 water, raw water, saline water,
 sea water, spring water, surface
 water, well water
 pollution 104f
 treatment 22, 46, 49, 51ff, 66, 87, 99, 124,
 126, 134, 139, 164, 173, 175f,
 179, 183ff
weathering (geo.) 18, 21, 50, 57, 83f, 90,
 128, 132
weight (body) 15, 27, 33, 61, 149, 150, 156
well (water), *see* water 23, 98, 119, 122

West Bengal (region, *see* India) 3, 14, 20f,
 35, 43, 57, 119
wheat, *see* cereals, plant(s) 17
woman/women, *see* female 25, 30, 37, 102,
 157, 159, 165
wood (preservation) 7f, 12, 16f, 62, 65, 78,
 122, 143, 151
workshop (soc.) 91, 93, 97, 105, 110ff, 133,
 172ff, 194

Xanthoparmelia farinose, *see* lichen(s) 114
Xanthoria elegans, *see* lichen(s) 9

yam, *see* vegetable(s) 145
yard (soc.), *see* garden 19, 63, 68, 99f, 123,
 141, 143, 146, 161, 195

zinc (Zn) 7f, 16, 18, 50, 77, 120, 130, 134,
 144, 146ff

Arsenic in the Environment

Book Series Editor: Jochen Bundschuh & Prosun Bhattacharya

ISSN: 1876-6218

Publisher: CRC Press/Balkema, Taylor & Francis Group

1. Natural Arsenic in Groundwaters of Latin America
   Jochen Bundschuh, M.A. Armienta, Peter Birkle, Prosun Bhattacharya,
   Jörg Matschullat, & A.B. Mukherjee
   2009
   ISBN: 978-0-415-40771-7 (Hbk)

2. The Global Arsenic Problem: Challenges for Safe Water Production
   Nalan Kabay, Jochen Bundschuh, Bruce Hendry, Marek Bryjak, Kazuharu Yoshizuka,
   Prosun Bhattacharya & Süer Anaç
   2010
   ISBN: 978-0-415-57521-8 (Hbk)

3. The Taiwan Crisis: a showcase of the global arsenic problem
   J.-S. Jean, J. Bundschuh, C.-J. Chen, H.-R. Guo, C.-W. Liu, T.-F. Lin & Y.-H. Chen
   2010
   ISBN: 978-0-415-58510-1 (Hbk)